C000001858

Nikola

Tesla

Prophet of the Modern Technological Age

(The Captivating Life of the Prophet of the Electronic Age)

Estelle Monk

Published By **Simon Dough**

Estelle Monk

All Rights Reserved

Nikola Tesla: Prophet of the Modern Technological Age (The Captivating Life of the Prophet of the Electronic Age)

ISBN 978-1-77485-572-0

No part of this guidebook shall be reproduced in any form without permission in writing from the publisher except in the case of brief quotations embodied in critical articles or reviews.

Legal & Disclaimer

The information contained in this ebook is not designed to replace or take the place of any form of medicine or professional medical advice. The information in this ebook has been provided for educational & entertainment purposes only.

The information contained in this book has been compiled from sources deemed reliable, and it is accurate to the best of the Author's knowledge; however, the Author cannot guarantee its accuracy and validity and cannot be held liable for any errors or omissions. Changes are periodically made to this book. You must consult your doctor or get professional medical advice before using any of the suggested remedies, techniques, or information in this book.

Upon using the information contained in this book, you agree to hold harmless the Author from and against any damages, costs, and expenses, including any legal fees potentially resulting from the application of any of the information provided by this guide. This disclaimer applies to any damages or injury caused by the use and application, whether directly or indirectly, of any advice or information presented, whether for breach of contract, tort, negligence, personal injury, criminal intent, or under any other cause of action.

You agree to accept all risks of using the information presented inside this book. You need to consult a professional medical practitioner in order to ensure you are both able and healthy enough to participate in this program.

Table of contents

Introduction

Electricity is, without doubt, an essential aspect of everyday life. When you turn on the lights or charge your phone, or turn on any electronic device, you are consuming the energy. In simple terms, your daily activities could not be achieved if there was no electricity. The man who was the source of that discovery Nikola Tesla, has been largely lost to history. In actual fact, his name isn't as recognized or well-known to the general public in comparison to his co-worker Thomas Edison or to other people who have made contributions to the science field.

Nikola Tesla, a Serbian-American engineer and physicist, is the person responsible for Alternating Current (AC) electricity. Tesla invented the first AC motor that could create and transmit electricity. This is similar to the one that is in use in all homes and buildings today. Tesla's contributions aren't just only limited to AC electricity. He was also a key influencer in the design of common devices like the radio fluorescent lighting, xray machines, and Radio Detection and Ranging (RADAR) among many others. Additionally

Tesla was the first to discover the magnetic field that rotates and invented an electrical circuit that was polyphase. The genius was and visionary. He often formulated thoughts in his head that he would later turn to actual ideas. In truth, Tesla has around 1200 patents registered with the US Patent Office that are issued under Nikola Tesla's signature.

Although he was regarded as an expert in his field however, he did not have the business acumen that many of his peers like Edison did. However the plethora of inventions he came up with were not able to translate into financial prosperity. In addition, his own mind was at its peak. In the final few months of his career Tesla would lie awake for long nights tackling scientific issues as well as mathematical formulas that he would juggle in his mind. As Icarus being too near to the Sun, Tesla was aiming for targets which were not realistic and ultimately resulted in his demise.

Despite his flaws and mistakes, Nikola Tesla is highly revered in the world of science. His achievements which are acknowledged by Tesla as well, certainly transformed the way that people are living today. This book offers an understanding of Tesla's life, the work of Tesla as well as the things he accomplished in his time and how he contributed to shaping the world of today.

Chapter 1: The Early Days Of Nikola Tesla

At midnight on the 10th of July in 1856, Nikola Tesla was born in Smiljan, Lika, in the region of Croatia. In the time of his birth it was part of the Austro-Hungarian Empire. Its father, Milutin Tesla, was an imposing but compassionate Serbian Orthodox priest and skilled poet of his own. The mother of his son, Djuka Mandic, though not educated, was a smart woman who loved to play with various electronic gadgets and developed household appliances to simplify the tasks of farming and home. Tesla's mom was a major factor in shaping his imaginative thinking. He also had four siblings: three sisters with the names of Marica, Angelina and Milka and a brother who was named Dane. Both brothers and he enjoyed building and doing things with each other. One of their favourite activities was waterwheels, with certain ideas that Tesla would later use as the foundation of his turbines with no blades later in his life. It was 1863 when Dane was killed in an horse-riding accident , which turned out to be a devastating event for "Niko" or "Niko" as he was often referred to in his home by family members. The family was profoundly affected and Niko was later disapproved of from his family, which led Niko to briefly run away from his home.

After the tragedy Tesla's father was elevated to the church in Gospic. Gospic. The family relocated a few miles away from the town, where his father was also able to begin instructing. Because Niko was already enrolled in classes at his home school and decided to continue his formal education at the Normal School. In 1870 in the year 1870, when Niko was just 14 and fourteen years old decided to go to the Higher Real Gymnasium in Karlovac (Karlstadt). He was taught mathematics and other languages in this school and excelled academically. One of the first signs of his intelligence was his ability to do integral calculus, which led his teachers to believe that they were cheating. He was also greatly inspired by Martin Sekulic, his physics teacher who demonstrated a myriad of ideas through the inventions were his. Due to his advanced understanding and determination it was possible for him to finish a four-year period in just three years. He graduated in 1873.

Prior to his graduation Tesla began to think of ways to think of an approach to inform him father that he didn't wish to be a priest. Tesla was enthralled with math and science and decided to study Engineering instead of being priest. But, his father was insistent and encouraged to allow him to enter the priesthood. He believed that engineering could be too demanding for a sick and fragile Niko. When he was in his final year at Karlovac the hospital, he was struck by fever while in the nearby marsh. This turned into malaria. The fever was cured within a few days, however his immune system became weaker. After his graduation, he was directed by his father to go back in Gospic for a few days and also to go for a long hunt to prevent the outbreak of cholera within the city. Tesla did, however, return home and eventually contracted cholera. He was in bed and nearly died from the disease. When he was sick and in bed, his father told him that if he recovered and remained healthy, he would let him to pursue engineering at the most prestigious institution. As time went on, when he was better and recovered, his family enrolled in to attend the Polytechnic School located at Graz, Austria.

It is interesting to note that what Niko as well as his entire family omitted to consider is the Army summons. He would be required to serve for at least three years with the army. In the time of this there was a significant war kicking off with Turkey. Turks and the nation was required to recruit new soldiers to the military. People who refused or were unable to sign up were penalized and imprisoned. According to the orders by Milutin, Niko went into the hills and hid for the duration of a year. The year helped him regain his health, while preparing for his college. In 1875, following the Military Frontier Authority forming a new organization, he returned back to Gospic and began his schooling the following year.

Tesla took classes in mechanical and electrical engineering in the Polytechnic School. In his first year there was a fracas in his class that made profound effects on his. His physics professor Professor Poeschl was the one who showed him and his students the Gramme Dynamo, which operates on direct current, and functions as a generator, as well as motor. After having a look at it, the professor suggested the brushes, also known as commutators, could be eliminated to increase the efficiency of the motor. Enthusiastic, Poeschl suggested to Tesla that it was similar to the idea of creating a continuous motion system, which up to the was a daunting task. In the following seven years Tesla was obsessed with his concept, convinced that alternating electrical currents was the answer to his dilemma. Alongside this continuous quest He also dedicated his time to study in order to convince his father that the decision to send his son to an engineering institute was not an error. He studied nine subjects that was more than the amount required and passed all the subjects in flying colours. As a result his health was again impacted. It was also not known to his father, he received notes from his professors telling him that he must be removed from his school because he did not wish for Tesla to pass away from exhaustion.

In his second season, Tesla made the decision to devote some of his time playing and attended much less classes than he took. He began to develop an interest in card games, and it became his method of relaxing. Apart from playing cards and playing, he also started playing billiards and chess. In the end, he began gambling. At some point his bank account was wiped clean which his father had gifted him to pay the fees at school. He was denied his scholarship and was kicked out of the school.

In the autumn in 1878, the student pulled aside his plans to study at his first year at the University of Prague and took the job of an engineering company within Maribor, Slovenia instead. The company paid him an annual salary of sixty florins as a result of his job. He was also awarded an extra bonus for his work. The hefty salary allowed Tesla to save money but still manage to live modestly. Despite his father's request to return to his to his home country, Tesla was able to stay in Maribor. The next year proved to be a test for Tesla after he was deported from Maribor due to a lack of a residence permits. Additionally, on April 17 1879, one month after returning home to Gospic, Milutin contracted a disease that killed him in the year he turned 60. In order to fulfill his father's wish to allow him to go back to school and reapply to the University of Prague using his income and the assistance by his father.

He spent a full year at the University and then quit due to a finances and inability to concentrate. He was so focused on finding out the details of his alternate theories that he slacked off on his studies.

With no funds, Tesla started to teaching apprenticeships, but discovered it was not a good fit. Tesla then sought help from his uncle Pavle to help him find work. At the suggestion of his uncle, Tesla moved to Hungary to work for a phone exchange run by Pavle's friend from the military, Ferenc Puskas. The relocation to Budapest was not just the beginning of his engineering career but also his breakthrough in the theory of alternating currents which has been his obsession for many years.

Chapter 2: Tesla's Beginnings And The

Alternating Current Systems

Nikola Tesla travelled to Hungary with such excitement and optimism. Alexander Graham Bell had just invented the telephone in America and Europe was opening up an open reception to it with the central station set to be built in Budapest the capital of Hungary. To his dismay, however, the exchange business Bell was expected to work had no jobs available. Furthermore, it was not possible to make one for him as the company was not yet in operation. Instead of the engineering job that he believed he'd be able to be offered, Tesla settled for a job as a draftsman in the Central Telegraph Office of the Hungarian Government, which had the emergence of a telephone within its area of responsibility. He earned a modest salary which barely enough to cover his basic needs, but he was desperate , and believed he must get a job or be starving.

Within a few months of starting his new job, Tesla displayed exceptional skills which attracted the interest of the company's Inspector-in Chief. This led to his move to a more senior position, in charge of developing calculations and estimates for new telephone systems. When the telephone exchange began operations, he was appointed the post as Chief Electrician. This was a huge relief for him. The young 25-year-old Tesla was the head of the business and initiated numerous improvements and innovations in telephone central station equipment. While working at the exchange for telephones Tesla was able to create his first invention, the amplifier also known as a telephone repeater, but the invention was not granted an invention patent. In the present, it is often referred to for its loud speakers. However, despite this accomplishment Tesla's main interest was on the alternating current theory that he was unable to resolve.

In addition to occasional activities with his friend Anthony Szigeti, Tesla devoted his time and energy to improve directly current machines in order to eliminate the commutator , and make use of the current alternating, without the need for bulky intermediaries. It got to the level where he had to take a nap in order to prove his theory was right. The zealousness caused his body and brain to fail and he suffered a severe nervous breakdown which could have claimed his life once more. With the assistance of an athlete Szigeti who forced Tesla to exercise and go out in the open and the determination to keep up the research he was conducting, Tesla slowly recovered from his illness.

In an outing together with Szigeti at the parks when he was reciting an excerpt from Goethe's Faust the poet stopped mid-sentence and experienced an epiphany on the issue that has been bothering him for years. With a stick and sandy material, Tesla drew a diagram of what is called the magnetic field that rotates. In his design it was a two-circuit circuit instead of the traditional single circuit that conducts electricity, thus creating two electric currents that were which were completely out of phase with one another. This resulted in the process of induction, a magnet would spin around across space, attracting electrons either of a the negative or positive charges.

The development of the magnetic field provided the solution to the issue that had afflicted Tesla for an extended period and gave him a new sense of optimism and a sense of. Tesla did not stop with the development of the model. He also created designs for motors, transformers and dynamos, and other components needed to construct the complete alternating current technology. Furthermore, he enhanced the effectiveness of the two-phase system by allowing it to operate with at least three alternating currents simultaneously. This would eventually become his famed multiphase system of power.

The telephone exchange which employed Tesla was eventually sold. Ferenc Puskas, who was the founder of the company, suggested Tesla to join the new Compagnie Continental Edison, which was designed to manufacture motors and dynamos as well as install lighting according to Edison's patents. Edison. Following the advice of his former employer to leave to go to Paris on April 18, 1882. After receiving a letter of recommendation from Puskas and his company, he was appointed an engineer as a junior for the company. He was employed at Ivry-sur-Seine. In the course of his employment the company, he was sent in Strasbourg, France for an assignment. He was tasked with fixing an indirect current lighting system that was damaged in a trial run. The system was installed by the German Railway. The government was in agreement with his work , however Tesla did not get the compensation promised when he completed the work. Believing he had been cheated, Tesla decided to terminate his contract at the company. Charles Batchellor, one of the company's administrators as well as manager of the work was adamant for Tesla to go to America. United States and work with his former acquaintance Thomas Alva Edison, who was the world-renowned Thomas Alva Edison. Tesla sold some of his books and other personal items, packed his bags and booked tickets from the United States to New York. But luck appeared to be in his favor when he found that his wallet that contained the tickets for his ship and train as well as his cash, and luggage was taken. The only thing he left was loose change that was enough to pay for the train ticket. When he got to the vessel, he told the incident to his fellow passengers. His crew were skeptical, but eventually allowed him to embark on the vessel even though nobody had claimed his space. Tesla was on a long voyage without a wardrobe or money leaving him unhappy and unhappy with the current twist of events. Amidst the chaos, there was a riot within the ship. He was threatened with drowning, but he was able to fight it off.

After what seemed like an interminable time, Tesla finally arrived in New York with nothing but the traces of his poems, a few of his belongings stolen and mathematical equations were he trying to solve with just the sum of four dollars in his pocket and a recommendation written by Charles Batchellor introducing him to Thomas Edison. In spite of his circumstances Tesla was optimistic about the future, as Tesla had reached the "Land of the Golden Promise". However, he will soon realize that not all promises are made to be kept.

Chapter 3: The Move From America: Tesla

Meets Thomas Edison

When he was 28 Nikola Tesla started a new chapter in his career. Tesla travelled across the country to New York City with the dream of meeting and working with Thomas Edison, who was the most famous electrical engineer of his day. It was the latter part of the 1870s that electricity was initially introduced into New York. The incandescent lamp Edison invented led to a booming demand for electricity. In the end, Edison's direct current power station located at Pearl Street in Manhattan, was becoming increasingly the monopoly.

In the Edison's office, Tesla was both excited and anxious to meet his idol. With the recommendation letter that was provided by Batchellor the engineer was able connect with Edison. In the discussion, Tesla went on to describe the type of engineering work he's completed, his polyphase power system, and his plans to develop an AC motor. Edison was a man who primarily used direct current, knew nothing about his AC system and didn't attempt to understand it. He laughed at the notion and straight-forwardly informed Tesla the there wasn't any hope for AC current. In simple terms the idea was that Tesla believed that AC electricity as an threat in comparison to DC power. Tesla and Edison clearly differed from one another - Edison was a believer in trial and error and believed that the effort and work put into the project makes the difference. Tesla relied on his own ideas and would create an idea in his head before taking action. In spite of his reservations about Tesla and the divergence between them, Edison still hired him because of Batchellor's recommendations about his work with the DC machine in Europe.

After a couple of weeks his appointment, Tesla was assigned by Edison to fix an issue with the steamship Oregon. The lighting dynamos were not working and the only solution to fix them was to pull them off and put in new ones. Edison considered this to be unacceptable since the date of sailing for the ship had already been pushed back and his company was placed in a sour spot. After Tesla came aboard, the engineer noticed an armature coil had burned out because of short circuits. With the help of team members, he took on the task of working on the issue and within an hour and a half, the machine was like new. The feat was astonished Edison and within the next few minutes, he was in up to a higher position, doing greater hands-on involvement.

In the course of his research, Tesla detected several ways to enhance the dynamos' structure so that its operation and efficiency could be increased. He approached Edison with a vision that would not just increase output, but also reduce the costs of operation. Edison recognized the benefits of Tesla's plan and offered his client $50,000, if they could achieve his goal. The huge sum was attractive to an utterly poor Tesla who was struggling to survive. Therefore, he set about developing his idea. He created 24 distinct types of dynamos, and successfully replaced fields with lengthy cores by smaller, more efficient ones. He also added automated control for the designs. After several months of effort the engineer was able to finish the project. Tesla later went to Edison to ask for the funds. In reply, Edison told him that Edison doesn't understand "American comedy". Incredulous that promises were discarded with such ease, Tesla immediately resigned from his position in 1885. Tesla did not receive payment for his inventions or designs.

In the time employed by Edison, Tesla has built an image in electrical and engineering circles. Following his departure in the year 2000, investors approached Tesla and offered to finance the creation of a lighting company in his name. Seeking an opportunity to earn money and implement his ideas then he accepted with them and Tesla Electric Light Company was created. However, the investors did not prefer to utilize this alternating current technology. Instead, they requested to create an arc lamp which could be used in factories and streets. In just a few months the designer was able to develop the design for the lamp and obtain patents. Green light issued to manufacture the lamp.

Tesla believed that it was the beginning of a career that would be successful for Tesla. Unfortunately, that was not the situation. In the course of development the developer was paid an extremely small amount of money . His agreement stipulated that he would be paid via shares. At the end, all that he was given were stock certificates. He was in fact forced out of the company he was employed in. When he attempted to convert these certificates into cash the company discovered that they were of no value.

Again stricken by financial problems, Tesla had to work as a day laborer, doing minor electrical repairs and ditch digging at a cost of $2 per day. He began questioning his capabilities and believed that his education was being ridiculed. In 1887, while on the course of one of his trench-digging sessions when he had a conversation between the foreman. After listening to Tesla's story and his thoughts the foreman was impressed and was able to introduce him to Alfred K. Brown, the director of the Western Union Telegraph Company, as well as his colleague Charles Peck, an attorney. In exchange for funding and backing, Tesla agreed on a fifty-fifty share of his patents. In no time, Tesla was able to construct a laboratory.

Chapter 4: Tesla's Rise To Fame

Nikola Tesla was able to create some of his ideas become reality in a relatively brief period. While he was at it his popularity within the electrical industry increased. In the months between November and December 1887, he'd filed for seven patents within the field of power transmission as well as a the polyphase alternating current motor. They cover a complete set of generators, motors and transmission lines, lighting and transformers. Tesla's designs were so unique that patents were granted easily. In the year 1888 Tesla had obtained over thirty patents on his ideas.

4.1 Collaboration with George Westinghouse

The results spread throughout the city and he got invited to present his work on transformers and motors that alternate current at the American Institute of Electrical Engineers at Columbia University. The lecture attracted the attention in George Westinghouse, head of Pittsburgh's Westinghouse Electrical Company and inventor of the air brakes for railroads. When he heard about the development by Tesla,

Westinghouse believed that it could solve his problems with power transmission over long distances. He visited Tesla as well as his lab and offered to purchase his patents for $60,000 which included 150 shares of Westinghouse Company. Westinghouse Company and $5,000 in cash. Apart from that, Westinghouse agreed to pay Tesla royalty in the amount of $2.50 per horsepower.

The technological breakthrough made by Tesla's inventions sparked an industrial conflict between the AC method of Westinghouse and direct current systems developed by Edison and was known as "Battle of the Currents". In an effort to change the public's opinion in Edison's favour, Edison launched a propaganda conflict on his AC system, asserting that it was a dangerous system. To demonstrate his argument, they employed a prisoner on execution row William Kemmler, as an illustration. Kemmler was electrocuted inside the chair with the Westinghouse generator, and died in an incredibly painful manner on the 6th of August in 1980. This incident was known as "Westinghousing".

Despite the bad press the situation was improving for Westinghouse. The company was awarded the contract to illuminate the world's first all-electric fair called The Chicago World's Fair. They beat the newly established General Electric Company by providing the most cost-effective and efficient alternating current technology. The fair's inaugural event was held on May 1st in 1893. It was a time of thousands of incandescent bulbs were created for the fair by Tesla, Westinghouse and 12 new alternating current generators. The exhibition's "Great Hall of Electricity" the creation as well as transmission AC power was on show for the visitors of the fair to observe, making obvious that electricity's future is in the alternating currents. Since then, more than 80 percent of all electrical device purchase in the United States were for AC power.

4.2 Niagara Falls Power Project Niagara Falls Power Project

Another idea of Tesla which was fulfilled thanks to Westinghouse was harnessing energy from Niagara Falls. As a young child, upon seeing photos of the falls, Tesla declared that he could harness the energy of the falls in the future. Westinghouse was awarded the contract to oversee the project and requested Tesla's help. The contract was the result of a contest that was unsuccessfully held through the Niagara Falls Commission, which was given the task of coordinating the project, but was unable to secure the backing of the experts they made proposals. The commissioner's head was Lord Kelvin, a well-known British scientist who initially disapproved of to the AC technology until he was at an event at the Chicago fair. When his opinions on the system changed Kelvin went to Westinghouse asking if they could use alternating current in order to channel power from Niagara Falls.

The time of formation caused a lot of stress for the investors in the venture, which include some of the most famous names in finance like Lord Rothschild, J.P. Morgan, W.K. Vanderbilt as well as John Jacob Astor. After a long and exhausting five years the project was close to its final stage. There were some doubts regarding the efficacy of the project, but they all vanished as the switch was switched on. At midnight on December 16,

1986 the first electricity was able to reach Buffalo.

Although the project was a huge success, Westinghouse was financially drained from the battle against General Electric,. J.P. Morgan was planning to take sole control of the country's hydroelectric power supply and to purchase Tesla's patents. In an attempt to save his business, Westinghouse requested Tesla to ignore the original contract , which would have gave the inventor a huge amount of royalty. In a gesture of gratitude to the one individual who was a believer in his work and his ideas, Tesla tore up his contract and saved the company. While it was a good decision, and even if he was a part of Westinghouse's success, Tesla was left with constant financial issues throughout his life.

Chapter 5: Higher Education

Following the recovery process, Tesla continued his studies at the Higher Real Gymnasium. When he was in High School, Tesla took physics. The professor showed electricity, and this sparked Tesla's interest in the subject. It initially appeared to him to be an unfathomable phenomenon. Tesla wanted to study it further and gain more knowledge. Tesla excelled academically. Some professors thought that he was cheating since his academic performance was so impressive. However, he was not cheating however, and he was able to finish his studies a year earlier than is typical.

While at the high school Telsa lived with an uncle and aunt. The family was very strict in its rules. The food was delicious, but food choices were limited since his aunt believed that Tesla was too fragile to be able to eat much more. Tesla had a frenzied appetite and was afflicted with. He also suffered from malaria during his stay in the area. But, he entertained himself by regularly taking on rats.

Following his graduation Tesla's parents invited Tesla to an expedition to shoot. A few days later, Tesla learned that cholera was prevalent in the vicinity of his home. He went back to Gospic and almost immediately the cholera virus was contracted. He was in bed for about nine months. He was repeatedly in danger of the possibility of dying. In the midst of Tesla's illness during a time of sadness over the state of his son's health, Milutin promised Tesla he could go to the top engineering college in the event that he could only recover.

The following calendar year Tesla almost was drafted to the Austo-Hungarian Army. But, he was able to avoid this by fleeing towards the south of Lika. In Lika, he was able to spend time in the wilderness taking in the mountain landscape. Later, he said that this adventure strengthened him mentally and physically. more resilient, which helped him recuperate from the illness.

After his time in nature, Tesla was able to attend the Austrian Polytechnic located in Graz, Austria. The school was chosen by his father because of its repute. Tesla was highly motivated to be successful. He worked from

3am until 11pm every day, striving to be the best. He was awarded top grades and passed 9 exams in a span of time, which was much more than the required. He also participated in extracurricular activities by establishing an Serbian culture club. The technical faculty's dean sent a note of praise to Milutin. It read "Your son is at the top of his grade." Despite his achievements Tesla's father didn't recognize his talents. The result left Tesla being very unhappy after all the effort to impress his father.

Tesla continued to devote his time to mechanics, physics as well as the study of mathematics. He would spend hours in libraries with the "veritable obsession" to work. In the library, he went through the entire writings of Voltaire. His teachers were impressed by his quality of achievement. However, they often didn't agree with his theories. If questioned, he went forward with his own thoughts.

In the course of his studies there were more letters sent to Milutin informing them of the fact about the fact that Tesla had been working to much. Tesla was later informed that his father was urged to remove him from the university. In fact, Tesla soon started to confront issues. For instance, he occasionally had outright disputes with his professors, if they disagreed with his assertions. Tesla also became involved in gambling, and eventually lost his scholarship as well as all his academic funding. Tesla did not have the right preparation for the next set of tests and had to drop out of school.

To hide from his family, Tesla had quit school, Tesla left the area and shut off contact with anyone who he was familiar with. Friends were concerned they believed he drowned in nearby Mur River. Actually, Tesla had moved to Maribor for a period of time.

He was employed as draftsman, and was able to relax by playing cards. When Milutin was informed of Tesla's whereabouts He visited him and begged him to come to his home. There is a possibility that, at the date, Tesla may have had the onset of a nervous breakdown. Tesla was later required to return to his to his home, courtesy of the police due to the fact that he was homeless.

In April 1879, Milutin died. The cause of his death is unknown however some speculate that the cause was stroke. In the years following, Tesla taught students in Gospic. Two uncles collaborated to raise funds for the purpose of ensuring that Tesla could go to the school in Prague. The first day of January, 1880 Tesla was admitted to Charles-Ferdinand University.

He was, however, missing the prerequisite subjects, Greek and Czech. The only option was to be able to attend classes on philosophy. He was not awarded a grades due to his absence. After his arrival, Tesla again suffered a "complete breakdown of nerves." However the patient recovered again thanks to the support of friends.

Early Career

Being unable to go to not being able to attend school, Tesla left Prague for being unable to find a job. Tesla moved in Budapest, Hungary with plans to join his employer, the Budapest Telephone Exchange. The moment his arrival, the company was not yet operational. He was temporarily employed in the Central Telegraph Office. After it was established that the Budapest Telephone Exchange was in operation, Tesla took on the responsibility of the chief electrician. In his position Tesla was instrumental in improving the equipment, and later claimed to have invented an amplifier for telephones or a repeater. But, there's no patent to support the claim.

After a time, Tesla's boss, Tivador Puskas helped him get a new job. Tesla moved from Paris in order to join the Continental Edison Company. The move was a bit overwhelming for Tesla. He began the strict routine of both work and personal life. In the early days he was at the cutting edge of a new field, helping install incandescent lighting in the city.

The position was a great help to Tesla discover more about the use to electrical engineering. Due to his expertise in engineering and Physics, he was noticed and was commissioned to design and develop superior versions of dynamos as well as motors. He was also required to help solve issues at other Edison utility companies being constructed around Europe. In one instance He was sent for Strassburg to work in the Strassburg area. In Strassburg, he also developed his inventions.

Make the move towards America

In 1884, one of the directors in The Continental Edison Company, Charles Batchelor had been transferred to US to oversee the manufacturing division - Edison Machine Works. In the course of his transfer he requested Tesla should be transferred back to America. United States too. This led to the transfer of Tesla into America. United States.

After Tesla came to America and returned, he was amazed by the new environment. He initially felt America to be more primitive. After Tesla arrived in America, he was employed at the New York's Lower East Side at Machine Works. The shop was packed with hundreds of employees including field engineers and machinists. Everyone was working on an enormous electric utility in New York City.

In his time with The Machine Works, Tesla met Edison on a couple of occasions. His autobiography Tesla recounts the first time he met Edison:

"The interaction with Edison was an unforgettable event throughout my lifetime. I was stunned by the man, who had no the advantages of his youth and a background in science achieved many things. I had taken classes in several languages, dabbled into art and literature, and had spent the best years in libraries , devouring everything from poetry to stuff...and was convinced that the bulk of my time had been wasted. However, it wasn't long for me to realize I had done the most beneficial choice I could have made."

The autobiography of his father, Tesla says that he gained Edison's trust by working on a tough task. A steamer that was fast for passenger use was malfunctioning, and Tesla set out to fix the issue. He discovered that the dynamos were in poor condition however, he was able to fix them. When Edison realized that Tesla had been working late into the night and fixed the issue and fixed the issue, he gave his strong endorsement to Tesla.

Tesla was later given a special project to tackle. At the time when arc lighting was a popular technology, it was also in use. It needed high-voltage power. This was not compatible with Edison's incandescent low-voltage system. This caused Tesla to be unable to contract with a few cities. Tesla had been slated to develop the solution using an arc lamp-based street light system. However, his plans weren't realized, perhaps because a different solution was found.

Tesla worked for The Machine Works for only six months, Tesla quit. It's not clear what led to this. Many believe that he was given an incentive for his work, but later was not given the reward. According to reports, the promise of a reward could be a prank as the facts included in the autobiography of his suggest the possibility of a real bonus was unlikely. In any case, he left at the beginning of 1885.

Career Changes and challenges

Shortly after shortly after leaving shortly after leaving the Machine Works, Tesla began developing his own patents to create an arc-lighting system. It could have been his designs that he created during his time employed by Edison. He collaborated with the same attorney employed by Edison to get his patents approved. The lawyer brought Tesla with Robert Lane and Benjamin Vail. The two businessmen were willing to contribute funds for Tesla's own utility and manufacturing company. The company was named Tesla Electric Light & Manufacturing company.

Through that, Tesla was able to obtain more patents and set up an lighting system that was installed in New Jersey. The media reported on the system, awed by the new features that are in Tesla's latest system.

Unfortunatelyfor Tesla, the investors showed no interest in some of his concepts. Particularly, they did intend to support his research with Alternating Current motors, or new kinds electric transmission devices. When Tesla was able to get the utilities running and the investors chose to fund only electricity utilities. They opted out of the manufacturing aspect and business. Then, they established an entirely new utility firm and walked away from Tesla and leaving him without cash.

Tesla's patents were linked to the company as they were exchanged in exchange for stock (stock which was then useless). Because of these situation, Tesla had to get through, working electrical repair jobs, and even as an excavating ditch. In reminiscing about that some decades after, Tesla said of 1886 that it was a period of struggle and "My high schooling in different disciplines of mechanics, science and literature appeared to me to be an act of mockery."

But, towards the end of 1886, Tesla's life began to improve. Tesla had came across Alfred Brown (a Western Union Superintendent) as well as Charles Peck (a New York Attorney). Both were experts in the establishment of companies and also with advertising inventions. They were impressed by Tesla's ideas and agreed to fund his research. They also would handle his patents. The result was Tesla Electric Company. Tesla Electric Company. Tesla was to make one-third of the revenue from patents, and the other third was to be reinvested into the development.

from Breakdown to Brainstorm

Following Tesla's massive "nervous breakdown" Tesla was awash of concepts for new inventions. Spending time picturing machines and devising new ideas give him a "mental state of happiness about as complete as I have ever known in life." During that time, ideas came to him in "an uninterrupted stream." Over that time period, he conceived of many ideas for motors and system modifications--things he would later develop, that became associated with his great successes.

Productivity and Productivity

At the time Tesla was employed by the telephone company while working for the telephone company, he worked in his inventions and found some encouragement from the people whom he worked with. Though he experienced occasional setbacks however, over time it was possible to develop a number of inventions that were both useful and significant. Some of them were gadgets:

Alternating Current

In the days of Tesla technology was on the edge in harnessing electric power. There was no way to be certain how the best method to achieve this goal. So, a rivalry was created. A lot of people think of Edison with electricity. In fact Edison was a proponent of his DC current. This method was more costly for long distances and could cause dangerous sparks.

However, Tesla promoted the Alternating Current that Edison declared to be dangerous.

To demonstrate the risks of the AC method, Edison would electrocute animals in front of crowds. He basically invented the electric chair that was later used by people. In reality, Tesla was offering the AC alternative as a safer and less costly option. However, even though Tesla tried to show the safety benefits of the AC alternative, Edison and his colleagues repeatedly tried to discredit the idea. The approach of Tesla prevailed and is now the main source of power in North America.

Electric Lights

Tesla himself didn't create light. He did however develop and utilize fluorescent bulbs for about 40 years ago before they were widely used in industry. He also developed some of the first signs made from neon for the World's Fair, when he used the glass tube and bent it into various forms.

The Tesla Coil

Tesla created his Tesla Coil around 1891, based on the notion that Earth itself could serve as a magnet in order to create electricity (now called electromagnetism). It could be transmitted by frequency and taken in by receivers. Tesla Coil Tesla Coil uses two coils that are primary and secondary. Each coil is equipped with an individual capacitor. This is a storage device for energy, much like the battery. The coils are connected in a spark gap which is the space in which sparks can be generated. This lets for the Tesla Coil to shoot electrical currents across distances.

Magnifying Transmitter

Tesla's Magnifying Transmitter was a prelude to other devices, and was designed to be the first step in electrical energy transmission via wireless. He believed that the completion of this invention would have significant and positive humane consequences. When he thought about it and writing about it, he said:

"The greatest benefit is derived from technological advancements that lead toward harmony and unity and my wireless transmitter is an example of this. With it, that human voices as well as appearance can be recreated all over the world and factories will be driven miles away from waterfalls that provide energy aerial machines are moved around the planet without a pause as the energy of sun will be is controlled to make rivers and lakes for the purpose of motivation and to transform of deserts with no water into fertile fields.

The introduction of phone, telegraphic and similar applications will instantly eliminate statics and other disturbances that currently impose a few limitations on the use of wireless."

In his attempts to design what he believed to be an essential device Tesla continued to exhaust himself almost causing another malfunction. This delayed the conclusion of the project. His autobiography detailed the research behind the Magnifying Transmitter in the following manner:

"It is an resonant transformer that has an intermediate in which the components, which are charged to an extremely high voltage with a huge surface area and are arranged on perfect enveloping surfaces that have huge curvature radii in order to ensure that no leaks be detected even when the conductor is not bare. It can be used at any frequency, from small number to several thousand cycles per second and can be used for the production of electric currents of massive volume and moderate pressure, or those with a lower amperageand huge electromotive force. The tension of the electric field is simply determined by the curvature of surfaces on which charged elements are placed and the size that the former."

If this talk on science isn't understood by readers on the average, then it's a sign of the genius of Tesla's mind that he was not just able to comprehend the subject but also wrote it down.

Automatons/Robotics

Based on his theories that external forces act on humans, Tesla conceived an idea to create automated mechanisms that could be used to complete tasks. Tesla designed robots that could carry out a variety of tasks. His initial ideas and work on this set the foundation for the field of robotics.

Remote Control

A gadget that many people use everyday The initial remote controller was invented and used by Tesla. In 1898, he showed the remote control of a model boat. The device was powered by large batteries as well as radio signals-controlled switches. They moved the propeller, rudder and running lights. This method was later employed by the military when they sought to create remote-controlled warfare. While not built on radio waves, remote controls are extensively used in every modern household for a variety of devices.

Electric Motor

Tesla invented a motor which was powered by magnetic fields that rotate. This invention could have helped reduce the dependence on fossil fuels and oil. But the power and potential for use of this invention vanished during the 1930 recession. It wasn't until a few the following years that it came to use in a variety of appliances and devices for the home. In the present, it has been used in vehicles bearing his name.

A Near Miss

Perhaps, it's even famous, Tesla nearly discovered X-Rays. It is believed that Tesla along with Mark Twin became good friends in the 1890's. While in Tesla's laboratory, Twain posed for a photo. It was among the first photographs to be illuminated by incandescent light. In the following days, Tesla invited Twain to the lab for a second photo. He as well as photographer Edward Ringwood Hewett used a Crookes tube. When they looked at the photograph, Tesla found splotches on the tube and believed it was damaged. In the following weeks Tesla learned that another scientist Wilhelm Rontigen having used Crookes tubes to create X-Rays. Tesla discovered that this was the mistake in his photograph of Twain and if he'd only noticed it in moment, the discovery of X-Rays might have been due to him, not Rontigen.

Shadowed by Rivals

One of the most tragic aspects of Tesla's career is the fact that despite his brilliant and imaginative thinking Tesla's work was frequently eclipsed by the work of his competitors. Other great minds often were able to beat him in the creation of his work. Their names are more frequently associated with success of their work. It took a while for Tesla to become more known in the field of his research. A lot of people are more familiar with the following people:

Guglielmo Marconi

Marconi was an Italian Marquis and an engineer in electrical engineering. He applied his expertise and engineering training in the forefront of developing the technology to transmit radio signals over long distances. He is credited with being the first person to design an electronic radio telegraph. In this sense, he is generally regarded as the original inventor of radio. Tesla was also working on the radio , and it is now acknowledged that he invented Radio first. Three things however made it possible for Marconi to be recognized as the first person to create the technology.

First, in the month of March 1895, Tesla's lab was destroyed by the fire. It halted all Tesla's research in the time, as notes and research was lost. Tesla also lost valuable time in the pursuit of discovery and innovation, since the lab had to be rebuilt the lab in another location. The second reason is that Tesla was, admittedly, off the mark in the field of long-distance radio transmission. He chose the wrong path of trying to utilize electrostatic induction to transmit radio signals. He believed that it was possible to alter the earth's "electrostatic equilibrium" to transmit signals over large distances. It is also thought that his research was slightly censored after Marconi was denied patents by the U.S. Patent Office incorrectly granted Marconi his patent. It is believed that the decision could have been in the hands of Marconi's financial supporters (Edison as well as Carnegie). This decision also enabled to the US government to not pay Tesla royalty.

Thomas Edison

Edison is famous for his numerous inventions. The most notable of these could include the bulb that lights up. He created the phonograph as well as the technology used to film motion images. The inventions completely changed the lives and the options for entertainment. It is probable that most modern societies utilize the technology that Edison invented each and every day. But, Edison was not the only one to come up with these inventions. Tesla did too.

In reality, Tesla and Edison were contemporaneous. They were pondering a lot of their ideas simultaneously. As an example both were working simultaneously to develop electricity-powered power transmission. Tesla used the concept of Alternating Currents that Edison was adamant about. Instead, he advocated a more simple direct-current system. Thus, he created the divide in between AC as well as DC.

In the same period, Tesla was also coming up with a variety of similar ideas that would eventually become Edison famous. He worked for a long time on a system was hoped to transmit voices images, voices, and even moving images. In this manner he was among first and perhaps the very first to come up with concepts that we see today in radio, phones or cell phones, as well as television.

Actually, his initial concepts are the basis of the modern mass communication system.

Despite the significance of Tesla's theories, some of his ideas was never realized. Additionally, the work was accomplished by him fell in obscurity with time for a while. Many scholars believe Edison's legacy was in no way due to any specific achievement he witnessed however, it is due to the sheer quantity of work he created. Edison was so prolific because Edison also invented a revolutionary method of innovation. Edison ran veritable invention factories.

In the course of his career, Edison had dozens of employees who took smaller roles for ingenuity and creativity. When he was enthralled from an concept, he'd often give it over to someone else who would determine how to implement it. For instance when he had the idea of creating an Kinetoscope (a moving camera) He enlisted his assistant (William Dickson, along with some others) to carry out experiments and design an initial prototype. This approach did not only permit more work to be accomplished quicker, it also resulted in Edison could maintain an enviable financial position. It also ensured that his employees were on the payroll and allowed him the money he required to keep developing his ideas.

The death of a forgotten Mind

In in contrast in contrast to Edison the work of Tesla was slower. The financial backers of Tesla were unhappy and some of them withdrew their support of Tesla. The work he did, while impressive, was, in general, less successful. While he worked until the end the course of his existence, he was struggling over his health. For instance, even later in his life, Tesla reported to mystically talking to the pigeons of New York City. Tesla was also unbalanced, financial stricken, and at home when he passed away in his room rented to him on January 7, 1943.

Unfortunately, a genius mindwith such potential, could never fully grasp or appreciate his potential. After the death of his father, his work, and even his name were pushed into the shadows. The legacy of his work was dimmed however, others' shined more.

In recent times, there has been a renewed curiosity about Tesla as well as his works. Many recognize the important contribution he made to the field of ingenuity and creativity which enabled the advancement of technology today. People also praise him to be a hero in the field of science, and also in historical books.

In the present, Tesla is being recognized more than ever before. For instance the name Tesla was chosen for the e-car that was the most innovative firm Tesla Motors. The company was established in the year 2000 by Martin Eberhard and Marc Tarpenning. Presently, Elon Musk is also involved in the company. The aim of the company is to create commercially accessible electric vehicles. They have had success. Recently the Tesla vehicle was taken into space. This is a fitting thing to do, since possibly it will kick off the development of a new type of ultra-long-distance communication.

In addition, Nikola Tesla is often thought of as a hero in the world of geek culture. Tesla is featured in comics online and films documentaries. There's even a movement to establish the creation of a Tesla museum that would truly honor his work. In this way, many years after his demise, with the very ideas he came up with to come up with, Tesla is being honored in ways that he wasn't truly honored in his lifetime.

Afterword

Modern science states that 'The sun was once the present, while the earth is present, and the moon is in the near future.' From an incandescent mass , we have come from, and now into a frozen mass , we are about to change. In the laws of nature. And quickly and irresistibly, we are attracted to our end.

Nikola Tesla

Many consider Tesla to be an historical underdog since his achievements are often masked by the achievements of his competitors. Yet, he may be considered an "lone Wolf" in his personal life , but in the technological and scientific world. Not common at the time, Tesla was never got married or had children. He also seemed to have some personal acquaintances or friends.

In his field, Tesla often worked alone or with only a handful of very few collaborators. Tesla accomplished many things and leave an impressive science-based legacy. One might ask but what he could had done differently to get more recognition. To answer this question, his work and life is a good way to find solutions that he left to follow. A few of these lessons offer advice what to do, and others suggest what you can do differently to achieve success.

What lessons can you take about Nikola Tesla's work and life?

Do Not Let Setbacks Set You Back

Through his career, Tesla faced many setbacks. The first was that his parents, and especially his father, did not seem to be favorable to his ambitions. His father would often dissuade Tesla from studying and his career plan. Then, Tesla faced the loss of his brother that was something he witnessed. The loss had a devastating impact the life of Tesla along with his loved ones. Tesla was unable to express his accomplishments due to the fear of obscuring the legacy and memory of his older brother. In the following years,

Tesla faced many injuries and serious illnesses and each one required time to recover from. When he graduated from college and then entered the professional world, things were not straightforward for Tesla. Sometimes, he was not treated with the respect he deserved and was forced to start all over again several times. Even events like the fire that erupted in his lab along with the loss of equipment he built put him behind. He wrote about those experiences:

"My project was thwarted by the laws of nature. The world was not ready for it. It was way ahead of its time. But the same rules will prevail and will make it a grand victory."

Every failure caused him to lose energy and time, Tesla did not give up. He always went back to where the opportunity was even when the components were damaged and broken. He put the pieces could be put back together and then built a new one, and moved forward. It's a valuable lesson learned from Tesla. Life isn't always smooth sailing, and the majority of people will experience at the very least a setback. This can happen to an player who is injured and is unable to play. This could also happen to an prospective medical student who fails a class that is crucial. Whatever the reason or how difficult it may be try not to let these failures stop you from achieving your goals. Only way to get to the goal of success is to use the things you've got, create plans to improve it and continue to try. Like Tesla did.

Anyone can conquer the hurdles of their own minds

In his autobiography, Tesla freely discusses the numerous bizarre psychological experiences as well as "nervous breakdowns" which he suffered through his lifetime. The same experiences could have hindered his progress. However, he refused to let these issues, experiences events, situations, and incidents stop him from the goals was his goal. He was able to continue to work even though those strategies included perhaps routines. The latest research suggests that routines could, in reality, be very important for people who suffer from mental health issues. In light of his own struggles and ongoing efforts towards the goal of success, Tesla himself writes:

"Can anyone be believed that such a an utterly hopeless body can be transformed into a man with astonishing determination and strength. Being able to work for 38 years nearly without interruption, and be healthy and vibrant in mind and body? This is what I experienced. A strong determination to live and carry on the work, as well as the help of a dedicated athlete and friend completed the miracle. My health was restored and along with it the vitality of my mind."

In fact, if you're suffering from psychosis, depression, or anxiety or another mental health issue don't let it hinder you from living your life and living your best life. With help and treatment the health of your body could improve, and balance out your mental health. Even if you continue struggle, that doesn't necessarily mean that those issues determine your life. A lot of people be able to live and work with signs of anxiety or depression. Don't let the hurdles in your mind prevent you from reaching your goals regardless of whether they're just as easy as self-doubt and the desire to be perfect. Additionally, as Tesla was open about his own personal experiences and experiences, you should be willing to seek any help or support you require, to ensure that you can be successful and overcome your challenges.

Consider thinking beyond what you perceive

In the time of Tesla when he was a young man, he came up with concepts that were a step ahead of the times. These ideas led to not just the inventions he came up with as well as the inventions that followed him. For instance the autobiography (written in 1919), Tesla was conveying ideas that could be to be all too familiar to those who read it who are reading this today:

"This invention was among several that I included in my "World System" of wireless transmission, which I set out to commercialize upon returning back to New York in 1900. Regarding the initial goals of my venture they were clearly stated in a technical document of the time, which I quote "The World System has resulted from the combination of several unique discoveries that the inventor made in the course of his long-running study and experimentation. It allows not only instantaneous as well as precise transmission of messages, signals or messages across the globe and also the interconnection between current telegraph, phone, and other signal stations , without any changes to their current equipment. With it such as this an individual subscriber to a phone could call and talk to any subscriber around the world. A simple receiver, no as large as a watch will allow the user to listen wherever in the world, whether on land or at sea to a message or music playing in an other location, no matter how far away."

Tesla was the one who came up with this whole system that he envisioned creating using Tesla's Tesla Transformer, Magnifying Transmitter, Tesla Wireless System, and many other gadgets.

But, it was not easy to bring his visions to fruition, particularly as there were challenges. Although some of Tesla's gadgets began to revolutionize communications throughout his lifetime and even helped the government communicate better across the oceans. Today, people might envision the idea of an Apple Watch when they read Tesla's visionary essay. In simple terms, Tesla envisioned a world filled with his innovative inventions that could function better than they ever did.

In essence, Tesla was thinking beyond the limitations of what he was able to see. He was able to think of ideas that others hadn't. Sometimes, it led to conflicts, because he discovered ideas and solutions, and argued their claims against the unbelievers. If the man had not pushed beyond the limitations of perception and information, he wouldn't have come up with some of the brilliant ideas that influence the world of today.

Any person can imagine beyond what they see today and that's what you should think about it to create the future you're not currently living in. Like you are able to make use of your imagination to come up with innovative ideas for technology and other fields, you can utilize it to get over your current limitations. Even if you feel that you are limited due to financial issues, a lack of support or other reason, you can look past these limitations and find ways to achieve success.

Collaboration with Others to Gain More Success

In his personal life, Tesla was a "lone one." He had only a handful of friends , and had no family members of his own. He would occasionally try working with other people however, it didn't seem to go well. When other people offered him jobs and he later quit, he often left in what could be a bad political decision. He resigned from multiple jobs and was frequently rejected by his financial supporters. His method of thinking was very different from Edison who saw more success with the spread of his discoveries in the field of science.

It's hard to say which greater accomplishment Tesla might have had with more support from his professional and social networks. Perhaps, if he like Edison had a real "invention factory," this would've been much easier to build after setbacks and also easier to complete the work faster. It was, in the end the slowness of his production that frequently led his financial supporters to withdraw from continuing their backing. While it's straightforward to blame Tesla for his failure to collaborate however, it must be acknowledged that it was not entirely his responsibility. A portion of it was due to conditions and his own mental health could have had a hand in it.

However, his style of work and achievements do provide a an indication that collaboration is essential to success. It's hard for individual to be successful completely on their own. People require the assistance of family and friends. They also require coaches, teachers and mentors. There is no great added achievement that comes from doing an entire task by yourself, rather you run a greater chance of failure and failure. Take pride in working with others and knowing that the most brilliant minds didn't succeed on their own.

Even if People do not Recognize Your Work, Keep Working

In his own lifetime in his autobiography Tesla acknowledged that his vast work and the strength of his ideas were not recognized in the world of science. In a statement of his knowledge the author wrote:

"The progress of the human race is dependent on ingenuity. It is the primary result of the creative brain. The ultimate goal of the brain is total control of the mind over the physical world and the synthesis of the natural forces to meet human needs. This is the challenging task of the of the inventor, who is often misunderstood, and not appreciated."

After Tesla's death , there was a growing recognition of his work. For instance, later in the same year, a few Marconi's patents were canceled since it was recognized that Tesla's inventions were first. It was also the case that Tesla's AC system was later adopted as the worldwide standard for transmission of power. Even with all these achievements, they were much too late for Tesla to enjoy personally the glory of these accomplishments. Even today, and in his life his recognition is restricted.

But, Tesla also recognized that his work was not only about recognition. Tesla truly wanted to make a difference in the world. Many creations he invented were designed to aid in humanitarian causes. There are reports that he refused royalty payments and money, since his work and the product he created were his primary objectives. Instead, he worked in not only in obscurity, however also financial ruin.

It's remarkable that the man did not stop working. He was always thinking and working. He created inventions that have helped the world. He left behind ideas which grew into something bigger for the ones who followed him. Recognition was not there however, the results of his work and the effects on humanity were .

The main lesson you can learn out of this experience is, even the fact that you may not get an award, recognition or even a large amount of money for your efforts If you are certain that it's a worthwhile and essential work, then keep at it. It doesn't matter if it's a scientific undertaking or helping others, an athlete or in any area you are in where success is determined not by the appreciation of other people, but the acknowledgement from yourself for your efforts to achieve a task well accomplished. Try like Tesla and try your best to complete an excellent job even if you are not acknowledged immediately or in the future. Maybe in a tiny way, you'll be able to leave the results of your efforts in the dust.

Chapter 6: Working Alongside Edison

Tesla was a young man of twenty-six with an idea Tesla was certain would alter the world. In a bid to share his idea Tesla began his career as an engineer with the Continental Edison Company, in Paris. He gained valuable experience in installing lighting systems within Paris, the French capital. His innate talent in engineering was noticed and he was subsequently requested to assist Edison's generators and Dynamos. He was also asked to assist in the troubleshooting of Edison's utility services throughout France as well as Germany.

He decided that his most important opportunity was at New York with the great man himself. He had a functional prototype, though not a complete one, in the early days. Tesla believed that it was likely that if Edison was offered better technological advances, Tesla would take it up the idea, and develop it with the help of his junior staff. The possibility of this opportunity fueled his desire to visit America and connect with the person who inspired him.

In the end, I took on the trip and, after loosing my ticket and money, as well as traversing a number of missiles that included the mutiny that cost me nearly lost my life, I finally arrived at these beautiful shores with just four dollars in my pockets.

Tesla was born in New York on the 6th of June, 1884.

Now a poverty-stricken, twenty-eight-year-old immigrant, he was still filled with dreams of success, in this strange new land where he knew no one. At six feet two inches tall, he spoke with a the weight of a Serbian accent and weighed just over one hundred forty pounds.

The decision to move to America was, in Tesla's eyes crucial. Tesla was dissatisfied with trying to find an AC motor made in Europe. The attempts he made toward Germany and France were unsuccessful.

Another thing that drove the ferocity of his pursue of Edison was a letter that he had in his pocket. This was an eloquent endorsement by one of Edison's European acquaintances, Charles Batchelor.

Dear Edison I have two amazing men, and you're one them. Another is the young man.

Batchelor was born Christmas Day, 1845, in Manchester, England. An experienced draftsman and machine operator, he was posted for a trip to Newark, New Jersey, in connection with his job employed by an industry that made textiles. This was where the man met Edison who also had an office in Newark.

In 1873, he was working with Edison in the process of inventing. A pragmatic man, he served as Edison's "hands" who utilized trial and error to refine their ideas. He was referred to as "Batch," the "chief experimenter'. Together with Edison came up with ideas for products that could be made in the areas of electrical lighting, telephone, telegraphy, and phonograph. Tesla had every right to be sure about the endorsement.

The end of dreams

Since the late 1870s New York had benefitted from electric power. With the backing from JP Morgan, the monetary giant of Wall Street, Edison installed his first power plant situated on Pearl Street, close to the financial quarter. The power station utilized direct current.

At the time the renowned inventor, thirty-five year old Thomas Alva Edison was already an American popular hero. Edison's invention of the incandescent light bulb was a hit with the world's population. Electric light bulbs and electricity are now in every home instead of gaslights. While this was a major advancement but there was an opportunity for improvement, which Tesla recognized.

As a brand-new source of power electricity was not appreciated by the general public and this created social barriers that needed to be resolved. Many people were scared of electricity. Many saw it as something that was mysterious.

For instance, someone in 1900 believed:

electricity would leak out of the outlet] to accumulate on the floor. Then you'd walk into the dripping electricity.

It caused the outbreak of fires spontaneously. Animals were suspicious of it as well, with horses rushing in fear through the streets when they felt an electric shock through their shoes.

Even with the initial difficulties, Edison was thrilled about the direction that electric power was heading.

In America, Tesla became even more excited to meet this star of of electricity.

I was ecstatic to witness my meeting of the marrow with Edison. Edison had changed the world through his incandescent lamp. I was elated to demonstrate my electric motor which was powered by alternating currents.

Edison was not as enthusiastic with the direct current infrastructure, having put a large amount of money into the direct infrastructure that is currently in use.

Tesla wandered through the streets of lower Manhattan in search of Edison's office to present his groundbreaking idea. He believed that the renowned inventor would be happy for the opportunity to share the secrets. But beneath the earth were 80,000 feet of copper, which was the first electric grid that was direct current that was the work of

Thomas Edison that lit homes and powered factories located at the bottom of Manhattan. Tesla's persistent pursuit of alternating current was always a source of irritation for Edison.

In spite of his reservations, Edison hired Tesla on the instantaneously. But, they were made of a different cloth. Edison was a practical person and inventor, and was also than a commercial thinker. He was determined to create things that worked and then sell them at profits. Tesla was motivated more by expanding his knowledge of the mystery of electricity and the ways in which its power could be used in a myriad of ways. Tesla was a highly educated engineer. He knew both the maths and theory, and was able to do lots of mental modeling and solving problems.

Edison's method was quite different. Along with 'Batch' Edison relied on trial and error to develop his progress. He was more comfortable working with "things" where you could observe the cause and impact immediately. He wasn't the most educated. He didn't even attend college but instead, began his career selling newspapers and sweets on trains when he was thirteen. It was a lucrative business earning $50 per week. Thomas later switched to Telegraphy, working during the night so that Thomas could experiment during the daytime.

Edison used to refer to Tesla as a poet of science', a term was a term he used to describe him as of ridicule, highlighting the lack of business savvy. When Tesla presented the characteristics of an AC motor in front of his boss, he received an uninformed response. He was just wasting his time. there was no hope for AC current.

In a bid to be noticed, Tesla worked for Edison all day, 7 days a week and redesigned his generators. Edison said to Tesla "I've had many hard-working staff, but you're the one to take your cake'.

Tesla stated that fifty thousand dollars would be paid from Edison as a reward when he achieved success in increasing the efficiency of the dynamos. This motivated Tesla to push forward. He threw everything in the ring in order to please the genius inventor. He desperately needed Edison's approval as an inventor, and also his business expertise.

I walked into my way into Edison machine rooms, where I worked on the design of DC motors and dynamos. My working hours were 10:30a.m until the time of 5:00 a.m. the following day. After I had completed the work I left to Edison to pay him and he then left [the roomat the end of the day.

Tesla believed that enough was enough, when Edison's reaction was

We don't have an American sense of humor Tesla. Tesla.

The argument led him to be dismissed from Tesla's Edison Corporation. With no money and trying to survive, Tesla was forced to take his pride in the air. His earnings dropped to just two dollars per day.

I went through a period of painful tears as well as hard work digging ditches to support the Edison underground cable.

Tesla's naivety in business, his blindness to science and his faith in the natural world often led to his financial difficulties. He was left to wonder what his next meal could come from. The heart of his was breaking. Edison was a snub rather than delight in his. He was all alone, separated from family members and acquaintances, and devastated. After two years living in America and nothing to demonstrate his time in America.

It is understandable that Edison was skeptical of Tesla's ideas. Thomas is famous for his extremely difficult and painful experience of trying to bring the idea to become a reality. For instance the Edison museum states that Thomas and his team of assistants tested over 6,000 plant materials for the perfect light bulb element. In the end, Carbonised Japanese bamboo was selected. It not only burned for 12000 hours however, it also allowed Edison to avoid lawsuits from other patent-protected materials that were also used by other inventors searching for the most affordable lightbulb for household use. Edison was perseverant and was able to see things through.

Even even if the AC technique was of technical merit however, he would need to alter his entire system to retrofit a set of adjustments to allow it to function. He was emotionally attached to the system he had installed, and was unwilling to consider the idea.

An extremely disillusioned Tesla abruptly went off on his own trying to build an alternative future for himself. This was a reckless move which would have cost Tesla much. Tesla believed that investors would finance his work because they were enthralled by the mystery of electricity, and not the more realistic cold-hard return from investment.

He was able to get the money, but the first venture failed.

A partnership of Tesla, Robert Lane and Benjamin Vale was formed in 1884. It was called the Tesla Electric Light and Manufacturing Company. Tesla's goal was to create an arc lighting system. It is an industry that was growing rapidly in the electrical light industry which is ideal to light outdoor areas. It took him a full year to complete his work. He was awarded a number of patents for the inclusion of an automatic adjustment system as well as a fail-switch, and better technology for dynamos within the system. In 1886, his system was lighting the streets as well as some manufacturing facilities located in Rahway, New Jersey.

Despite the advancements in technology however, investors showed no enthusiasm for Tesla's plans for electric motors and transmission equipment. The tight market conditions created by two rivals made it hard to turn profits. Investors' eyes were attracted by the idea of establishing an electrical utility, which would operate in distribution and generation. In the fall in 1886 they established their own business Union County Electric Light & Manufacturing Company which would be the end of Tesla's venture. Tesla was again left in debt and lost control over his patents that were given for sale to Lane as well as Vale instead of stocks.

The Rebirth of the Dream

Despite the abrupt twist of events, Edison was not wrong to be enthusiastic over his alternating current. There were many disadvantages to Edison's system, and problems that were that were so urgent that resolving them ultimately resulted in the triumph of the alternating current.

Firstly, despite being so celebrated, Edison's current exhibited a major problem-- transmitting it over longer distances was fraught with difficulty. The wires "lost" power and needed to be amplified using other circuitry.

Second, changing or reducing DC voltages required complex circuits.

The third reason is the volume of copper wire required to provide reliable power for long distances was huge with wires as strong as an arm's length to be able to support it.

In the beginning of electric power, the public had less expectations of service quality and reliability and had no issue with the proliferation of power stations located one mile apart. As a contrast, the first transmissions that were long distance in AC throughout America were about 14 miles (22 km).

As Edison's primary goal was to power light bulbs and small motors, direct current was the best solution for urban regions. The distances between them led to the system's breakdown. AC wasn't subject to the same limitations since it could increase and decrease the voltage in order to address the issue of range.

But, as a functioning AC motor was not in existence at the moment, it would be an uphill battle to obtain funding to implement Tesla's distribution system.

Despite everything not going according as planned, Tesla was still keen to create his AC motor. And thanks to the assistance of a group that included investors Arthur Brown and Charles Peck on Liberty Street, in 1886 Tesla was able to create his own lab just a few blocks to Edison's offices.

It was in this place seven years after the first time he had an idea and began to work on a solid prototype for his motor. Tesla created his AC motor by feeding several out-of-phase currents through it to produce a magnetic field. Tesla immediately patented his idea and Peck made the patents available to a businessman named George Westinghouse. Along with an electric motor Tesla developed a total transmission to use AC power that is operating until today.

On May 16th in the year 1888. Tesla gave an audacious lectureentitled 'An entirely new system of alternation current transformers and motors' in front of members of the American Institute of Electrical Engineers in New York.

The topic I am privileged to bring to your attention is a new motor that I am certain will soon prove the ability to adapt to alternating currents.

Tesla took advantage of his achievements that was now acknowledged and recognized as a part of electrical engineering's establishment. From 1892 until 1894, he went to become vice-president for the American Institute of Electrical Engineers that was the predecessor of the current worldwide institute (IEEE).

In the following five years, 22 patents for a wide range of devices that rely on alternating current including generators, motors transition lines, transformers and many more, were granted by Nikola Tesla. These patents have the highest commercial worth in the years since the introduction of phones, but very few investors were aware of their potential. One person who could be a potential investor was the Pittsburgh industrialist George Westinghouse.

Chapter 7: Inventional Work

Tesla was granted a lab in Manhattan at the address 89 Liberty Street. He could explore innovative ideas and improve existing concepts. His initial focus was on generators, electric motors along with similar gadgets. In the nick of time (in 1887) the inventor was able to design an induction motor running with Alternating Current. This system of power gained popularity because it could be used for long distance in transmission and also high-voltage power.

Tesla's latest motor operated using a polyphase current. It created a rotating magnetic field that turned the motor. It also permitted self-starting without any commutator. This reduced the possibility of sparking. Also, it reduced the requirement for maintenance and servicing. The first patent for electric motors was issued in the month of May 1888. Peck and Brown decided to make public the motor. They achieved this through independent testingthat confirmed the improvement in its functionality. Press releases were then issued along with copies of the patent.

In the early part of 1888, there were arrangements made with an organization called the Westinghouse Company to manufacture Tesla's motors, but on a larger scale. The plan was not without its challenges. There were some issues. Tesla motors were built in low frequency electric currents however, they were not compatible with the conventional designs for Westinghouse. Westinghouse apparatus. It was necessary to make adaptations.

Beginning in the year 1889 Tesla began working again from New York at his laboratory. Tesla was primarily focused on the design of high-frequency machines. This was an extremely challenging job because it was an innovative area. He initially rejected a technique that later realized would work. He also tried to develop the simplest method for creating electrical oscillations. In the early days, Tesla was inspired by Kelvin's work. Kelvin. Tesla was able to make rapid advances and developed the coil that could generate sparks that were as big as five inches. As per Tesla, "Since my early announcement of the invention, it is now in widespread use and revolutionized several areas. However, a brighter future lies ahead." In the future, the device could produce discharges that are as high as 100 feet and could generate large currents around the world.

In 1890, Tesla carried out a lab experiment using high-frequency currents in order to demonstrate that "an electric field with sufficient strength could be generated in the room to illuminate the electrodeless vacuum tubes." The invention was deemed a achievement and was received with admiration by the general public. When his accomplishments were recognized, Tesla struggled with the acknowledgment. Tesla was determined to concentrate only on the work he was doing.

Apart from that, throughout the 1890's Tesla was extremely productive. Tesla invented electric oscillators, meters, better lighting and experimented with X-Rays and demonstrated radio communications by navigating a boat around the pool (two years prior to the time that Guglielmo Marconi showed radio communication capabilities).

The year was 1891. Tesla together with Westinghouse was the lighting provider to The World's Columbian Exposition in Chicago. They also collaborated together with General Electric to place AC generators close to Niagara Falls, which was the first power station.

In the same time frame, Tesla conceived an idea to build a bigger machine that could accomplish more than anything else. However, in 1895, the laboratory of Tesla was destroyed in a fire. This set back his work and he needed have to build his laboratory before it was possible to resume work. He was able to resume work and continue to create machines that could generate ever-increasing voltages of electricity. To carry on the work, Tesla went to Colorado to work for a full year. At the end of his time Tesla came up with the "Magnifying Transmitter," which Tesla continued to work with and perfect over the course of time.

In the years following his return back to New York, he got financial support through J.P. Morgan. The aim was to construct an international network of communications with a tower located at Wardenclyffe. In the end, the funds had run out and Morgan was too shocked and irritated by Tesla's schemes to provide any additional funds. Tesla was once more constrained by financial limitations.

III

A Troubled Mind

I don't think there's any excitement that could fill the human heart in the same way as that experienced by the inventor when his creations of the brain advancing to the point of... these feelings make you forget sleep, food and friends, as well as love for everything.

Nikola Tesla

In his reflections on his childhood and his career growth, Tesla wrote,

"Our first ventures are intuitive, promptings of imagination that are vivid and uncontrolled. As we age we become more rational, and we become more systematic and able to design. However, those initial impulses while not always productive however, are of the highest value at the moment and could influence our lives. In fact, I am convinced that if I had recognized and nurtured them and not suppressed them I would have added a lot of value to the legacy I left for the rest of humanity. It wasn't until I reached the age of manhood did I discover the fact that I had been the inventor."

Strange Experiences

Tesla's autobiography, filled with information regarding his personal life exposes another reason why his achievements in life could be delayed. He reveals that as young the author "suffered from a niggling illness that resulted from seeing images frequently accompanied by powerful flashes of light that interfered with the view of actual objects and disrupted my thinking and actions." He also reveals that his ability to visualize objects or scenes he had before seen. These images were usually triggered by hearing the words, however the person would struggle to discern what he was seeing from what was actually happening.

Tesla states that these visions led to "great anxiety and pain." According to reports, he visited psychologists and physicians who could not provide a full explanation of what he experienced. Tesla believes his experience might be distinct, but it could also be something that he was predisposed to, as his brother was also suffering from the same issue. The theory he came up with was that the images resulted from an impulse action within the brain's retina "under intense excitation." Tesla does not believe that they were "hallucinations like those that occur in the minds of people suffering from mental illness and anxiety." The argument he uses in this regard is "for the rest of my life, I was in a normal state and was composed."

While the experience was a bit traumatic however, it did give Tesla the idea to make films: "it should be able to project onto the screen the image of the object that one envisions and show it in a way that is apparent." This type of advancement will revolutionize the way we interact with each other. I'm convinced that this dream is achievable and likely to be realized."

But, Tesla would try to stop the images from being distressing by focusing his thoughts to something else he'd witnessed. It was difficult since Tesla was not a fan of the world, and often did not have enough things to remember. The solution he came up with was usually ineffective or loses effectiveness due to the use of. The method he found beneficial was to instead go into his mind and then move through the mind of his.

While Tesla was able to occasionally remove the images out of his head with "willful determination," he was never in control of those flashes that he occasionally observed. They usually happened when the situation was "dangerous or stressful circumstance" or when Tesla was "greatly delighted." Sometimes it was possible to be able to see that the atmosphere around him suffocated with fires. The the intensity of these visions continued to grow.

When he was 25 at the time, he went on an adventure in shooting in the air and freshness energized him. The next night, he experienced the sensation that his brain was on the fire. He noticed a bright light that resembled a small sun. He applied cold compresses on his head all night. It took a while for the light flashes to fade, and it took three weeks before they were completely gone. The photographer never returned to another excursion following the incident.

Following that it was a while before he would see these flashes of light for instance, when an idea that was new occurred to him. But Tesla observed that they were less dramatic and interesting. The autobiography of Tesla describes how Tesla writes of closing his eyes to see "flakes that were green" light, then "systems of closely separated lines" with different colors as well as glowing light. But, seeing such images at night would often help him sleep and the absence of the images indicated a sleepless night.

In his autobiography Tesla writes about inaccurate beliefs and obsessive thoughts, cognitive distortions, compulsive behaviors, and aversions. Tesla writes about his childhood as a child, wanting to fly and believing that maybe the possibility of flying was. However, he also had numerous "dislikes and behaviors." He was averse to abhorrence to earrings, particularly pearls. However, other pieces of jewelry drew his attention. He was awed by hairstyles of others. The appearance of a peach would make him sick. He would track his steps every time walking and would systematically analyze the cubic volume of the containers. All completed actions had to be divided in three parts, even even if it took hours.

Beyond these bizarre behavior and beliefs, Tesla was also described as being "oppressed by the thoughts of suffering from death and life as well as religious anxiety. I was influenced by superstitious belief systems and was constantly worried about the evil spirit ghosts, ogres and ghosts and other evil creatures of the night."

Telsa tells us in his autobiography, that after having read the book, he was engaged in some kind of self-control that he was previously lacking. He was capable of exercising the will of his self. But, he soon was a bit too confident in his self-control abilities. In the end, the man began to engage in self-destructive actions, such as smoking cigarettes and gambling, and eventually became addicted. Over time the addict was able to end his addiction and quit all of his vices.

Mental Breakdown

writing over his breakdown which was experienced while in University, Tesla says: "What I felt during that time of illness is beyond all imagination." Tesla stated that his senses were always alert and so powerful that he could feel an alarm clock ticking from a room away. Tesla also described being overwhelmed by sounds, sights motions, movements, and fluctuations in pressure. He often felt his heart beating faster while his body moved in "twitches and shaking."

In later years He also experienced other episodes of mental symptoms in times of stress and work. In an autobiography of his, he writes about having been so stressed by an assignment that he began seeing visions of his past life. Even when he had these visions, he was still working in a state of discord.

In addition, he began to believe that he was gifted with supernatural or psychic abilities that enabled him to anticipate future events and even communicate with his mother in a short time after her passing. The scientific mind of his was able to see arguments against this, however he believed that humans are nothing more than automatonsthat are manipulated by forces outside. He wrote in his autobiography of experiencing a certain type of "cosmic" discomfort that he experienced when he or someone else was affected in the same way.

Strengths and Assets

particularly his troubled mind. Tesla was also thought of as a hard-working and persevering. When he wrote about this, he stated, "I am credited with being among the hardest workers, and maybe it is, as I believe that thought is the same as labor as I've devoted to it for the majority of my time. However, if you consider work as a specific deadline for completing a task following a set of rules that I might be the most inactive of idlers. Each effort that is compelled to perform requires an investment of life-energy. I have never had to pay such a price. Instead I have prospered from my ideas."

In his autobiography Tesla describes how he faced mental and physical exhaustion. He wrote that he rarely went on vacation, but when he was over-worked, which led to the build-up of "some poisonous chemical" Tesla was prone to "sink into a sluggish state that lasted for up to a half hour or a minute."

In the morning at night, he was capable of working with a new perspective and an ability to overcome any hurdles that could have previously stopped him from progressing. At times, he would be absent from work for months or even weeks and then come back to the task to finish it.

Additionally, Tesla was generally quite predisposed to exaggerating his capabilities. For instance in his autobiography Tesla claimed to have stomach skin "like the crocodile's" and the capacity to consume cobble-stones. He also recounts a story of throwing rocks at a trout who had jumped out of the water, and hit the rock just right, to press the trout against the river rock, cutting it into two. According to him, his uncle was witness to thisand caused him to be "scared to the point of losing his mind."

The controversial view points

In addition to having controversial experiences and behaviors, Tesla also believed in the use of stimulants to boost his productivity at work. Tesla believed that we should "exercise moderately and manage our appetites and desires to every direction." He also believed that those not able to do so, and may have suffered health issues or even death as a consequence of addiction were "assisting nature" by observing the principle which was "survival of the strongest."

Tesla believed that his ability to manage his "appetites" allowed him to remain "young in mind and body." The autobiography he wrote relates his ability to stop himself from a chilly fall even at 59. He also relates visiting an eye doctor regularly and having great vision after 60. In addition, he says he isn't being overweight between 35 and 60. This means that the clothes still perfect fit him even after all these years.

New Perspectives

Today, researchers are aware that, as Tesla lived, studied while working, Tesla may have been struggling with mental health problems that were not recognized in the early days. There is a belief that Tesla suffered from compulsive behaviour and obsessions that could today be classified as obsessive-compulsive disorder.

In the majority of cases, Tesla was seemingly able to channel his obsessional energy into his own creative sources of invention and innovation. The creativity and productivity could have greatly influenced his mental health problems. In fact, the analytical mind developed by his obsession could have helped him with his job. His records indicate that he was often working for long hours and nights, with no sleep. He was often keenly analyzing technology that was accepted by others as it was. This led him to search for more efficient and better ways to design the technology. This also led to conversations with people who were around him, such as his professors, which proved challenging throughout his studies.

The condition of his mental health also led to some unusual symptoms, which were problematic for Tesla. He had a routine life that included a regular schedule. He worked every day from 9am until 6pm. Dinner was served every evening at exactly 8:10pm. Tesla always went to the same place and was required service by the exact waiter every evening. Tesla also had a number of physical ailments and, among them his tics, he believed that he had to curl his toes at least 100 times a evening.

Additionally, Tesla is reported to have suffered from sickness for periods that he would even experience visions. It is believed that a portion of these ideas were used to create inventions or solutions for the various technical issues he discovered. Given the period of time it was viewed by many as a behaviour as simply a result of genius, simple peculiarities and irks, or perhaps something like the term "mad inventor."

It is now believed that Tesla might have been afflicted with an autism-like condition known as high-functioning. Researchers are still looking into the genetic connection between autism and the ability to be genius. It is widely known that families with children who are autistic also have a higher percentage of children with a high IQ. These links are linked to cases of savantism.

For Tesla the diagnosis is not conclusive several years later following his demise. It is well-known that he displayed some signs of behavior. However, autism can also be related to social deficits that Tesla didn't necessarily display. Tesla was definitely notorious for his isolation, since he would often lock himself away in his lab to work for long durations of time.

However, Tesla was considered to be elegant and charming in his social interactions. He was not a very social person however there were some close friendships. Tesla was also a man with a broad spectrum of interests that went beyond the sciences and technologies. He also loved languages, music and philosophical thought. These aren't as than those of people with autism.

It's impossible to know in certainty what Tesla's mental health profile or diagnosis might be in the event that he had been assessed thoroughly with modern methods of psychological assessment, and whether Tesla truly met the criteria to diagnose OCD or the Autism Spectrum Disorder. But for many, seeing Tesla as a result of this assessment provides the understanding of his actions, and provides a sense of confidence for their own capabilities to overcome the past mental health issues and still perform actions of amazing skill.

IV

A Forgotten Mind

In the 21st century, robots will take over the place that slaves was employed in the ancient civilisation.

Nikola Tesla

Tesla was among the most innovative inventors and innovators throughout history. He was devoted to his work due to the belief the following: "An inventor's endeavor is essential to the survival of. When he harnesses force to improve devices or offers new comforts or eases, he's adding to our security existence. He's also better equipped than the average person to defend himself from danger as he is alert and highly resourceful."

Inspiration for invention

looking back at his life that was a source of inspiration, Tesla writes in his autobiography of his first invention. In the first invent Tesla created an apparatus as well as a method. The incident occurred when, as young one of his classmates was fishing with a hook and tackle. All of them went to catch frogs, However, Tesla was left behind after a heated argument about the kid. Tesla did not have had a hook before, but the idea was to build one. He succeeded and then set out to catch Frogs. He initially was unable to catch frogs but eventually, he managed to catch them by hanging the bait before an frog on a tree stump. He was able to capture two frogs, and then taught others how to do it.

Successes and setbacks

The next invention by esla was designed to "harness the energy of nature in the service of mankind." He decided to make use of May-Bugs/June-Bugs to drive an rotation. The device worked. But another kid appeared and began eating live insects right in front Tesla. This was so alarming that Tesla that he resigned from the project and refused to do any work with insects again.

Tesla began to turn his attention to disassembling and reassembling clocks. He was pretty adept in disassembling clocks, however, he wasn't as proficient in getting them back together. The clocks belonged to his grandfather who was not pleased with the situation , and stopped Tesla from using other clocks. It took him thirty years later that he tried to get back into working with clockwork.

Following that, Tesla started trying to create a pop-gun using the hollow tube with a piston and hemp plugs. To trigger the gun Tesla had to push on his stomach against the piston, then push it back using both hands. The air was compressed between the plugs, and elevated the temperature. This would remove an individual hemp plug. The device was successful, however, it led to damaged window panes. His family was against using it again.

Following the popularity with the gun, and being dissuaded from continuing to work using it Tesla began carving swords from furniture pieces. This was inspired by his love for the poets and their heroes. With his swords, he would attack corn-stalks and damaged the crops. Tesla's mother would beat Tesla for this.

According to reports, all of these events occurred prior to the age of six. In its second year of The Real Gymnasium, Tesla became obsessed with creating an apparatus that could be moved by constant air pressure. Tesla had a difficult time in putting the device together but was finally capable of creating an instrument. The device consisted of a cylindrical unit that could freely turn over two bearings.

Practical Ideas

After recovering from cholera after spending a year in the wilderness, Tesla conceived of some ideas. He conceived of an invention which could help transport packages and letters across vast lakes. He thought this could be achieved using submarine tubes, which are cylindrical containers. He was of the opinion that the pumping plant could push to move water into the tube thereby moving the spheres. Another option was to build the equator with a ring. It could flounder free and be set in the direction of a spinning. He believed that this would enable rapid travel. Another idea was to harness the energy generated by rotation of terrestrial bodies.

Visionary Capabilities

Tesla noted his ability to visualize things as an aid to his imaginative mind. At the age of 17, when his focus was focusing on invention it was possible to think clearly without drawing models or even experiments. He was able to imagine things as true in his mind which enabled him to devise a fresh method to translate his ideas into reality. Tesla believed that visualization was the most efficient since the process of building the idea to reality can lead someone to get caught up in the project only to lose themselves within the particulars of the creation process. He believed that in the long run, the quality would be diminished. In his notes on his method, he said:

"My way of working is different. I don't jump into work. If I have some idea I begin right away building it in my mind. I modify the structure or design, then make changes and even use the device in my mind...In this way, I are able to quickly create and perfect an idea without altering any aspect. If I've gone as far as to incorporate into the device every possibility of improvement I can imagine and find no flaws and then I can put into the final creation of my brain...Invariably the device functions exactly as I imagined it would and the test is exactly as I intended it to. In the 20 years since I invented it there hasn't been any exception."

Tesla has also credited his long experiences with visions as enhancing his ability to observe and determining causality and effects. He was able to apply these skills in a way that was automatic, which helped him in his research and ingenuous projects.

Chapter 8: The War Of The Currents

Westinghouse was a gamble with the Serbian inventor due to his experience working by Edison. He invited him to join the company and also provided him with a lab in which it was possible to perfect his invention which could fetch millions if it was successful.

The last quarter of the nineteenth century when railroads became the main artery of the country there were a few Americans were incredibly wealthy. One of them was 22-year-old George Westinghouse, who had developed the air brake for steam trains that significantly increased the security of the mainline.

With the experience of successfully making inventions available to the market His eyes were set focused on the future. And for him, it was the lucrative, growing electrical industry.

Seeing Tesla's solution, Westinghouse thought he had found his man, the gateway to opening the immensely lucrative game-changer--alternating current.

While visiting Tesla's lab, the inventor offered the entire patents that underpin his AC system for a staggering $1 million.

To sweeten the deal more, Westinghouse proposed that for every horsepower produced by a device that uses the technology that is patent-pending, Tesla would be paid two dollars in royalty.

In awe and admiration, the other inventors watched in amazement, while the vibrant Serbian man was pushed towards fame and success.

It seemed like a perfect fit, Westinghouse and his business know-how and financing to assist the inventor navigate through the business world. Tesla was a genius in technology and innovative insight to resolve the engineering problems to make the profits.

In the direction of Pittsburgh, Westinghouse and Tesla started building Serb's dynamos and transformers, as well as motors. They were necessary to allow long-distance electric transmission. Even though Tesla had a backing however, financial problems would soon befall Tesla's plan.

Edison had already invested an enormous amount of capital to his venture. Westinghouse did not, and the expenses were staggering. Investors began to be nervous about the company's overextension. Tesla was told that if you wanted your AC motor to be successful the company would need to forfeit all royalty payments. In fact, if he had complied with the terms of the agreement, it led to the company being forced to go out of business. In defiance of any professional advice from a lawyer, Tesla was in agreement, and not considering negotiating an alternative payment plan.

With his financial backing and an AC system being introduced, Tesla was more of an apprehension for Edison.

The war gets underway

The War of the Currents, as it came to be known was the fight to decide how people would power home and their businesses into the near future. The size of the market was enormous. One side is Thomas Edison's direct-current system, On the other, the alternating currents coming from Nikola Tesla and George Westinghouse.

Each side claimed their method as more efficient and perhaps , more important to the public, safer than the other. Tesla will eventually prevail in the coming years.

Edison who was aware his AC generators had a tendency to spread across the nation more quickly then the DC alternative, decided that he needed to take desperate strategies. The result was launching an entire propaganda campaign in order to make his opponents look bad.

Every death that was resulted from AC was extensively reported. Edison declared:

As sure like death Westinghouse can kill customers within six months following the time the installation of the system regardless of size. The company's plans don't worry me in the slightest.

My personal goal is to ban all use of alternating currents. They're as ineffective as they can be dangerous.

The war became extremely brutal as it raged on. There were two major examples. Let's take a look at the first.

In 1887 an Edison fan proposed that Edison's AC system could be employed for live demonstrations in order to show that it is able to kill human. At the time there was a discussion concerning capital punishment. The saga of hangings that were not properly executed led to some concern and a committee was established to investigate alternatives in 1886.

An idea was devised to execute criminals from the New York's Auburn State Prison using electricity. It revealed how strongly Edison would like to destroy his foe, George Westinghouse. There was no pause in the fight. The rivalry between them was truly electric as they attempted to get more lucrative electricity contracts.

What made this bizarre notion get its start? It was 1881 and Buffalo dentist and part-time inventor Alfred Southwick, had read about a tragic incident in which drunken man died nearly immediately after touching an electric generator. To test the idea that electricity could provide an acceptable solution for stray dogs, he killed a number of dogs from The Buffalo Society of Prevention Cruelty for animals (SPCA). Then, he published research suggesting that the device could be a plausible alternative to hanging, and was then asked to join the committee. He dubbed the chair an 'electric chair".

On November 8th 1887 Alfred Southwick sent Edison a letter detailing his findings, and an odd request for advice on how to conduct electrocution on humans.

Initially astonished, Edison wrote back that the idea was repulsive and he'd:

Join hands with us in our efforts to eliminate capital punishment.

Southwick nevertheless, he wrote another letter one month later. He received a different response. an entirely different response.

The best equipment for this is the dynamo-electric machines that use intermittent currents. The most efficient of these are referred to as "alternating machines', which were manufactured in the United States through George Westinghouse.

This was yet another of Edison's flims attempts to demonstrate the Westinghouse AC model was a fast and effective killer, thus making it clear that his direct current model was more secure and gaining a larger market share.

Greed gave birth to the famous execution method which was used by New York as its machine of death.

Edison himself delivered a gut-wrenching demonstration of electrocution in front of reporters. He put water on an aluminum sheet and attached the sheet to an AC source of power, and then let a dog lick down the water. The dog fell to the ground dead when its mouth was in contact with the electric.

In order to ensure that Westinghouse's AC was utilized, Edison funded secret work to get an electrical engineer to swiftly create a functional device. The head of the capital punishment committee, Frederick Peterson, a neurologist who was involved in a few of the dog electrocutions , engaged a the consultant Harold Brown to finalise the set-up.

William Kemmler's execution

On the 5th day of December 1888, Brown began an experiment in Edison's West Orange laboratory. The press was invited. Also, doctor lawyers and members from the Medico-Legal Society who had also contributed to the decision-making process of the way in which the apparatus would function. Edison and the chairperson of the commission on death penalty were watching.

Brown utilized alternating current to conduct testing on animals bigger than human, which included a deformed horses and four calves all of them executed using seven hundred 50 volts AC. Afterward Brown, researchers from the Medico-Legal Society recommended the use of between one thousand and 1500 volts AC to execute human beings. Newspapers reported that the voltage that was used to power lines running across streetcars in American cities was more than double.

The first person to participate was supposed to be William Francis Kemmler (9 May 1890 - 6 August 1890). He was a slum rat that were Buffalo, New York. A vegetable peddler who was alcoholic, the man was convicted of murdering Matilda "Tillie" Ziegler the wife of his common law to death using the hatchet.

Following an incident, the man been to his neighbour and admitted the crime.

Westinghouse and its employees, who do not want their technology to be used for this bizarre goal, has backed several legal proceedings. One of them claimed that the electric chair was in violation of eighteenth amendment's prohibition against harsh and unusual punishment. Kemmler was hired as the best attorney money can afford, and took the case to before the Supreme Court. Alas, the death sentence was not overturned, despite the hefty one-hundred-thousand-dollar legal bill.

At sunrise on the 6th August, 1890 inside his prison cell, the convicted killer donned his suit and put on an elegant pair of polished shoes. He was taken to an empty wooden chair, made of oak by the warden in a packed room. Kemmler stated, in media reports:

Ladies, I wish you all every success in the world. I'm convinced that I'll go to a wonderful location. The newspapers have been expressing lots of things which isn't the truth. This is all I can say.

Then, with Kemmler strapped to the chair, electrodes were secured to Kemmler's skull by the warden who said goodbye to William after signalling the assistant to turn the switch.

The absurdity began to unfold. After 17 minutes of electrocution, at an infinity distance, two medical professionals concluded that Kemmler had passed away. The electric current was shut off. The body fell down.

The silence grew in the room. It was quickly interrupted by someone shouting:

Wonderful God He is alive!

It was evident that Kemmler was breathing while his heart continued beat. As he tried to salvage the situation, another person shouted:

Make sure you turn on the current!

After a long and painful wait while the generator was charging at two thousand volts, rather than 1,000 now travelled through the body of Kemmler. Kemmler began to bleed as his hair started to sing as the horrible smell of burning flesh made a few witnesses look like they were giggling.

It took four minutes for the man newspaper now calls "the poor wretch" to end his life. The process caused the body to reach an alarming temperature which took several days to get cool. A witness said that his spinal cord ignited.

Tesla commented:

The death penalty is not just cruel and inhuman, but also unneeded as a part of the modern world of civilisation

It is not surprising that Westinghouse was equally horrified, and was believed to have stated:

They could have had a better job done with an Axe.

Southwick was in attendance and had a slightly different view:

This is the culmination the ten years of work and study! We are living in a more advanced society from today.

Edison was equally optimistic about the events and put the barbarism on'some incompetence' that was caused by the exuberance. In the future Edison offered suggestions on how to get a good result:

I am of the opinion that when the next person is seated to be executed the death penalty is likely to be executed immediately... It is better to place hands inside water jars, and then let the flow be activated there.

Then eventually, the War of the Currents began to slow down. The new safety procedures helped save the reputation of safety for the AC cable network on the streets. Fusions reduced competitiveness between the companies. Edison Electric itself merged with its primary AC competitor, Thomson-Houston, forming General Electric in 1892. The fight was then focused in securing profitable contracts instead of the prior battle for publicity.

Topsy the elephant's execution Topsy the elephant

But, there was to be a incident in 1903 in the following year and involving the electrocution of Topsy the elephant using the use of alternating current.

The Edison manufacturing company was present to record the moment. (Their short, silent black and white documentary, titled "Electrocuting an Elephant," can be viewed on YouTube in the present, if are able to stomach it.)

While it is possible to be claimed that it was only an issue of the press, given that it was the War of the Currents had officially ended for many years and the presence of Edison's staff fueled reports that the rivalry maybe still alive and well.

Born to Southeast Asia in 1875, Topsy was transported to America shortly after. For the next 25 years, prior to the time she arrived at NYC's Coney Islands Sea Lion Amusement Park in 1902, she was made to join an elephant group that was part of the Forepaugh Circus. At this point she gained the reputation of being difficult to manage.

A fatal accident that killed a viewer in 1902 caused her to be sold and relocation into Coney Island, where Topsy was involved in a number of dangers and was thought to be to be unmanageable. A slurred handler who used a pitchfork to brutally take control of the animal may be a contributing factor.

Determined to turn the demise of the kid-friendly facility into an advantage, publicity-hungry owners, Thompson & Dundy, decided hanging the public would draw the attention of a large audience and generate an impressive amount of money. The SPCA came in to protest the violence, and the concept of hanging was discarded.

The group then suggested that electrocution, strangulation, or poisoning could be the best option It would also be a less formal event that only invited guests.

Topsy Standing between journalists and other onlookers she refused to go over the bridge across the lagoon until she reached the point where she was to be shot. She was eventually secured in the area in the spot she was standing.

The carrots she was fed were laced by potassium cyanide. A rope was tied around her neck, and then attached to winches that tightened it. Electrodes were attached to her feet. When the switch was turned and six thousand volts were flowed through her huge volume. She was more fortunate than Kemmler because it took only 10 seconds before she tripped over and passed out.

Chapter 9: The Columbian Exposition In

Chicago

Let's go back a bit earlier, back to the time the days when distribution of electricity was the main issue of war.

It was the year 1893 and hostilities were about to get to boiling point at the Columbian Exposition in Chicago.

Significantly to Edison the event was going to be the first fair lit by electricity. Its financing, organization and management of the fair was handled by a variety of prominent professional, civic and business leaders from across the United States, important potential clients for those working who work in the electrical business.

The scope of this event was impressive. Fourteen magnificent structures were designed by famous architects of the day. It was a huge six-hundred and ninety acres, roughly the same as 11 million tennis courts. It had nearly two hundred mostly neoclassical, top-quality temporary structures. There were lagoons and canals. More than 27 million people from 46 countries took part in the exhibition throughout its six months of operation. The person who secured the contract to bring the event to life would soon be the most well-known name in the world.

In a bid to impress the attendees and organizers, the newly-merged Edison Company and the Thompson Houston Company, General Electric was formed on the 15th April 1892, made an offer of one million dollars to win the auction.

Not wanting to be outdone, Westinghouse also put in an idea that was much less expensive, and with a amount of half-a-million dollars.

It was likely that the lower bid get accepted by the group that was coordinating the event.

Westinghouse eager to show the alternating current technology, was now working on a massive initiative in Chicago to show off his latest technology. The massive AC generators were used to supply all the fair's electrical power.

The company's nose was knocked out of place In a retaliatory stance, General Electric refused to offer Westinghouse the Edison lightbulbs. They also asked for a judge's intervention and ban Westinghouse from using one-piece lightbulbs that are of any type at the trade show.

In a bid to not be outdone, Westinghouse saved the day by frantically constructing an e-stopper lamp that was two pieces right in time for the fair. The lamp's design did not violate Edison's patents.

After years of being a victim, Tesla had a chance to become a hero of the legend.

The 1st day of May in 1893, a hundred thousand spectators were gathered in the stunning venue. It was the perfect moment to allow Tesla and Westinghouse to shine.

As night turned to day and the president Grover Cleveland pulled a lever. The entire fair was lit up with light, stunning white tube lighting and multi-coloured lights that directed the way. It was the most spectacular display of electric lighting that anyone had ever witnessed.

The thrill of the spectacle was overwhelming and awed by the beauty and spectacle of electricity, those who attended as well as the rest of the world realized it was the Tesla-Westinghouse technology which gave it the life it needed.

Tesla eager to retaliate against the anti-science campaign Edison had been waging against him, and by displaying impressive showmanship and awe-inspiring demonstrations to demonstrate the power of his inventions.

Tesla was determined to amaze the crowd of many important investors, as well as an array of engineers. Tesla was aware of the need to gain support, just like his colleagues in the field in the moment.

The audience was not going to be impressed by complicated equations and well-written journal articles. Telling rather than showing was the key to the fair.

At the moment, Tesla had begun to investigate possible ways to transmit electricity using wires.

In his hands was a basic tubular made from glass filled with special gasses. The astonished crowd watched the tube move through space, the tube emitting a pleasant glow. Incandescent lamps that had filaments were standard at this time. However, Tesla's lightbulb that had no filament, and no wire connection or to a battery was a total originality. Electricity from the ambient, or an electric field, was sufficient to ignite the gas. The invention of his father was the prelude to the fluorescent tubes that we have today.

Ambient electricity was generated through a different Tesla invention known as the Tesla coil. It was able to produce extremely high voltages through making use of a weak AC voltage and then constructing it within the coil. It produced a huge quantity of power. One spark could unleash all the energy instantly.

The skin effect is a way of understanding where electricity would flow through the epidermis before descending to the earth instead of attacking the internal organs of his body and organs, he would extend his hand and generate sparks.

His body could flow many thousands of volts electricity that would leave Tesla absolutely unharmed by the process. The public was amazed. The procedure was not carefully tested in the lab and was perceived as fascinating, even if it was a it was a risky choice.

He was dressed in a tuxedo with a white tie, an over-sized hat. He would place his hand on a wire that would emit electricity across his body, causing the most beautiful light shower.

The view was extraordinary and people were watching at awe and amazement.

One journalist described Tesla as "acting as of a true magician as a showman , with his electric prowess between magician science and the business.'

Tesla completed his three-hour talk with the statement that electricity should become the future and a'servant to humanity's needs'.

The energy and power that results from it will ensure that humanity make huge leaps. The amazing possibilities open up to our minds, increase our expectations and make our hearts swell with extreme joy.

Tesla was motivated by his fascination and awe of the universe of nature and science instead of money. He was even more determined to harness this power and make use of it to alleviate the suffering of humanity and their toil.

Following the expository

Events of the exhibition left forever in the hearts and minds of Americans. There was a feeling renewal throughout the entire city. There was a sense of rebirth in the city. Great Chicago Fire from the 8th through the 10th of

October 1871, killed around three hundred people, burned three-point the area of three square miles (9 km2)) in the town down to the ground and left over one hundred thousand residents without a place to live.

Chicago evolved into a futuristic metropolis moving forward as a living representation of Tesla's innovative ideas and vision.

The fervor of his success was palpable. Tesla was conscious of his childhood dreams of harnessing the power of nature. The majestic Niagara Falls.

From the very beginning there were other scientists interested in exploring the possibility and an international group headed by a renowned British mathematical engineer and physicist Baron Kelvin, turned its focus to the same issue.

The idea was first suggested, Kelvin, in a cable to commission members from other agencies included the following warnings:

I hope you don't make the huge error of alternating current.

Fortunately for Tesla Kelvin's view, it was completely altered following his participation at the Chicago fair and his first-hand knowledge of the AC system operating. Even though he personally still preferred direct current and direct current, the confident Kelvin quickly signed the huge contract to supply generators to Westinghouse Electric.

The Niagara project

It was to be a major technical project which required three five-thousand-horsepower generators, the largest ever built at the time. There would be a myriad of problems to solve before the installation could be completed. The only thing is that Tesla was confident that it was going to work when it was put in place.

The issues led to heated discussions between Westinghouse technicians and Tesla especially over the issue of how to determine the ideal operating frequency.

The technical aspects have been worked out. All that is left to do is to wait for the button to be put in.

It was, thankfully, successful.

The huge generators came into life, powered by huge shafts linked to water turbines which were pounded by massive Upper Niagara water flow.

The generator's current was increased through a series of transformers that stepped it up. 22 thousand volts of electricity was transmitted across long distances. Later, it was decreased to power motors as well as small local communities. The system also follow Tesla's blueprints.

It was reported in the Niagara Falls Gazette proclaimed:

Today, the tramsof this city are operating on fraudulent power. The Falls will have to work hard to earn their livelihood.

Tesla was proud that his dream had been realized, stated:

Imagine my delight when, thirty decades later I viewed my boyhood plans being executed at Niagara and marveled at the inexplicable mysteries that was the human mind.

In the case of transmissions over long distances direct current was a technology which was defeated. Alternating current was the winner.

Prior to Tesla was invented, there had to be thousands of tiny power stations on every mile. After Tesla and a single source of power, Niagara Falls, the entire northeast could be lit and powered.

After the initial installation proved to be as a huge success, in the following years, the number of generators located at Niagara was upgraded to 10. When the century of the 21st century began New Yorkers would gain of the electricity lines that stretched for a staggering three hundred and sixty miles from the city.

The battle of the currents that lasted more than a decade ended, which meant that Tesla won the war.

Westinghouse's financial troubles

The situation was not positive for Westinghouse. Despite the technical achievements of the alternating current, it came at a price. Company resources were declining and stretched beyond their limits. Many times, when it was performing loss-making operations to expand market share the Tesla's financials was a risk. For this reason that Tesla's royalty payment was not able to be paid.

It's unclear the source of this notion. Could Tesla be coerced or even manipulated, as Edison was thousand dollars or did he come to the conclusion by himself? There isn't any evidence that can be found in the Westinghouse archives, just an unsigned memorandum.

Perhaps the confident, flighty and innovative Tesla was looking forward to his next adventure now that his boyhood dream was accomplished and he was ready for what was after?

Due to the discoveries he discovered in his own studies, he became convinced that his next invention would be millionaire.

I had already spotted enough to grasp the possibility of energy being transferred and received with no wires.

My work with Westinghouse not being required I recommenced my work in a laboratory located on Grand Street. I began immediately to development of high-frequency machines based on the successes of Niagara.

Drawing on his prior experiences the inventor created a plan. He intended to use his inventions to improve people's lives and free them from the burdens of daily life, whether at the office as well as at home.

Due to the successes of the exposition and Niagara the mysterious foreigner has captured the attention of all who heard about his incredible electrical gadgets. The public was eager to learn more.

Franklin Chester, in The Citizen on the 22nd of August 1897, wrote about Tesla:

As far as appearance is concerned, nobody will be able to gaze at him without feeling the force of his presence... He has cheekbones that are prominent and high and a sign for the Slav. Eyes are blue, deep set, and his eyes are like fireballs... When he speaks you are captivated. It is possible that you do not know what he's talking about and yet he charms the listener... He is in a bizarre mysterious world, reaching out to harness new power in order to acquire new knowledge.

Return to square one

In awe of enjoying his Niagara victory, Tesla chose to show in Delmonico's Restaurant, which is the most costly restaurant within New York and was led to his table elegantly dressed, wearing an elegant derby cap, a classy waistcoat, and wearing stylish glove made of white leather.

Diners were reported as:

He was extremely concerned about his health and drank one glass of whiskey every day. He said it would increase his lifespan to hundred fifty years old.'

While dining at the Waldorf within the Palm Room, he was so well-known that people came just to see the inventor. His meticulously recorded timekeeping made it easy to anticipate his actions. Around seven in the night, he'd sit down for dinner.

Tesla became a household name and was a modern Edison. There were many stories about his warmth, charm and humour were abounding. He was fluent in several languages and a lover of poetry, paired with his brainy mind He was awe-inspiring.

His sponge-like mind was eager to impart details on any subject. Reporters reported that he had something interesting and informative to say, regardless of the topic. He was a modern day Leonardo di Vinci - a scientist and a man of the arts.

Looking for a second financial backer to fund his innovations, Tesla networked with the most influential movers and shakers of the time such as wealthy politicians, millionaires, and famous people.

One of his close friends from his circle of friends in the New York elite were writer Mark Twain who said:

I just saw the sketch and description of an electric device recently invented by Mr. Tesla that will change the entire electric industry in the world.

Another close friend of mine was John Jacob Astor IV (13 July 1864 15 April 1912) Astor was an American businessman who was a real estate developer, inventor, investor, and writer, and even a Lieutenant-Colonel during the Spanish-American War. His fame would come to an end with his death in the time of the sinking of the Titanic in the early morning hours of April 15th, 1912, becoming one of the richest individuals to die in the tragedy, valued at more than two billion dollars today in modern money.

William K Vanderbilt received fifty-five million dollars (around one-six billion dollars in the present) through his father. He also was the manager of his family's investment in railroad. In 1879, following his acquisition the Barnum's Great Roman Hippodrome located located on railroad land near Madison Square Park, he named the facility Madison Square Gardens.

Robert Underwood Johnson, a former diplomat, writer and the editor of the well-known Century magazine Century was among Tesla's closest associates. He went on to become the US ambassador to Italy.

Johnson's wife was enthralled by Tesla and was able to take care of his needs, including cooking him meals and helping him stay focused on his work.

My dear Nikki I beg you to leave those millionaires as well as Fifth Avenue for some simple pleasures, not distinguished through a huge weakness

Katherine Johnson

Tesla wasn't just the world-renowned inventor. He was the most interesting bachelor in the entire city of New York.

Though he never got married, he was always interested in women and they were drawn to him. Many sought his attention. He was charismatic and charming. Two of his most famous snatchers include Flora Dodge, an attractive socialite, as well as Sarah

Bernhardt, the famous French actor Sarah Bernhardt.

When Tesla was sitting with a fellow in a café, Sarah dropped her handkerchief in front of Tesla. Nikola was quick to the tissue and grabbed it before bringing it back to her. In a low bow, he handed the tissue back to her telling her: 'Mademoiselle. Your handkerchief. He didn't notice her smile as Nikola had returned to his chair to continue his conversation with his companion about his research into wireless power transmission.

In relation to his concerns about how the influence of romance on his creative abilities, Tesla commented:

I don't think there are many amazing inventions created by married couples. I think the writer and musician should get married. They are inspired and lead to better results. An inventor, however, has such an powerful a personality, with many aspects that is wild and passionate that, in committing himself to a woman the inventor would be willing to give up everything and thus eliminate everything from his area of expertise. It's sad that he would also be a pity.

Despite his phenomenal mental capacity, Tesla was plagued with numerous phobias and it was apparent that he suffered from what we could consider to be an obnoxious-compulsive disorder.

Certain of these worries put limitations on his ability to build relationships with people particularly women. He could not bear to look at earrings, and he was uncomfortable at the idea to touch hair, and preferring to stay clear of the scent of perfume. He was not a fan of shaking hands too.

While using his free and inventive mind to come up with an endless supply of patents, rigid routines and obsessive thoughts were always present. Every time he could, everything that he did was to be divided by three, like the number of rooms in his hotel. In his youth, he completed 27 laps every morning. Before entering a structure, he would stroll through the neighborhood three times.

All of these factors together were likely to decrease the likelihood of developing a relationship. At the age of forty, he said that he'd decided to be consciously unable to express his sexuality.

All women, shakers, and movers that he met were put in a second place the inventions he came up with.

Although he was a great humanist Tesla was a man with a lot of problems with people, and even though he was passionate about love but there was no room or place in his world for romantic love.

I'm not sure there's any excitement like that experienced by the inventor who can see his creations of the brain taking shape to achieve greatness. These emotions can make one forget about sleep, food and friends, as well as love for everything. It's not easy to be lonely.

Exploring high-frequency power

Always interested, along to his AC energy technology Tesla became interested in investigating high-frequency electricity, and was experimenting in the 1890s.

Prior to that, in 1873 in England Mathematical scientist James Clark Maxwell had proven that light is electromagnetic radiation that formed when electricity was operating at a very high frequency.

With another invention known as using a different invention, the Tesla coil (which continues to be used in the present) He set off on the journey to explore the new, undiscovered world. Its ability to boost voltages at high frequencies Tesla could send one of his radio waves.

In a bid to demonstrate his coil's potential, potential financiers were invited by his lab for exclusive , late-night demonstrations of science.

It is impossible to know what might happen when they entered Tesla's lab. Since electricity was an invention that it was frequently difficult for the observer to comprehend what could be revealed. Wires were heated within the palms of hands, and the lamps in their hands were lit when Tesla transferred thousands of Volts through the guests' bodies.

One of them included Mark Twain, always keen to play guinea pig.

Thunder is great. It is amazing. However, it is lightning that is the culprit this day.

Electricity captivated Mark Twain, and with Tesla always seeking out new ways to use electricity, Twain was a frequent visit to Tesla's laboratory. There were numerous photographs taken of the author as a gracious participant in the proceedings.

One demonstration was involving the mechanical oscillating machine Tesla was engaged in in the lab. It was extremely vibrating when it released its high frequency.

Tesla was interested in seeing whether there was a therapeutic benefit from exposure to oscillations. Twain was an enthusiastic participant. According to some reports, Twain had digestive issues and Nikola tried to ease the symptoms.

The machine was powered up and Twain took to the oscillator. After a few minutes of the machine, he declared that it was enjoyable and he felt an increase in energy and energy.

Many times Tesla demanded that the person remove himself from the machine however, the experience was so enjoyable that he would not let it go until the nature decided to take over. The oscillation device of Tesla was productive in the way of laxatives. Twain had to rush to the bathroom at an incredible speed.

Additional developments were made from the Tesla laboratory, initially neon light, followed by fluorescent as well as the first photographs of x-rays. Amazing as they were these discoveries soon faded as in 1890, Tesla made light without wired connections.

He considered it to be an early signal to test the possibility of wireless power, something that he had hoped for since his youth.

It was to me the first time I could prove the idea that energy was being transferred in an airspace... I am now certain that the realization of this idea isn't too long away.

News traveled quickly and, in 1892 Tesla was asked to share his findings from his experiments with high frequency to European scientists and engineers.

A bit more like science fiction than science the people at London in London and Paris were astonished by how much Tesla could achieve using electricity and lighting.

In his magic hand, he introduced an amazing new possibility: radio.

I'd like to say some words about the thought that is in my thoughts and worries. The wellbeing of everyone. This is the transmission of information and power, without wires. I am increasingly convinced of the viability of this plan.

The race for radio's attention was set to start.

The year was 1888. German scientist Heinrich Hertz His name, which is associated with the frequency unit Hertz discovered radio waves that were previously predicted in Maxwell's equations. He also demonstrated the existence of light as a form of electromagnetic waves.

The invention of the first radio transmitter as well as the first radio receiver Hertz and his apparatus demonstrated that electrical signals could be created in one area and observed in another without any between. It was demonstrated that currents of high frequency emit electromagnetic signals or radio waves to space. He did not imagine a practical application for his research.

This is of no value whatsoever... It's an experiment to prove that Maestro Maxwell was correct, there are these mysterious electromagnetic waves are not visible to our naked eyes. However, they exist.

But other scientists, like Marconi and Tesla might have a sense of the significance of the phenomenon even if it represented an enormous leap in the realm of imagination.

When he was conducting the experiments in England, Tesla encountered, Sir William Crookes. Crookes was an British physicist and chemist who was a student at the Royal College of Chemistry in London and was involved in the spectroscopy.

Crookes was a fervent believer in mysticism. This, through his research on radiometry, showed that both concepts involved discovery of forces previously unknown. He believed that humans could be attuned to high-frequency brain waves that could enable psychic communication, some were skeptical, believing Crookes was being manipulated by fraudulent mediums. The scientist Victor Stenger wrote that the tests were not controlled properly and that:

The belief in the supernatural was blinded him to the scheming in his psychic subject.

Tesla was skeptical too, but he began to be more open the concept after having one of his captivating nighttime visions.

I saw clouds carrying beautiful images of stunning beauty. One was looking at me with love and then slowly assumed the appearance of my mother... At that moment I felt a certainty, that cannot be expressed in words was awoken in me, the fact that my mom had recently passed away, and that was the truth.

Following this dream Tesla believed that the two of them were on the identical frequency.

Another groundbreaking real-world invention could result from his extra-terrestrial experiences.

Relocating in New York in 1893, Tesla disappeared into his lab located on South Fifth Avenue, banishing his social circle in order to focus solely only on the work he was doing.

The inclination led him to realize the possibility that Tesla coils that are tuned to resonate with the exact frequency could send and receive radio signals that were clear.

Tuning is the primary factor in every radio and TV transmission in my lab... I can put in my palms an instrument that was tuned to my body, and gather power wherever in the room with no physical connection. I've had a tendency to disappoint my guests by these amazing effects. At times, I'd emit fireballs shooting from my head, and then run an electric motor through my fingers for six or eight hours.

In Tesla's lab, it could have been like a kind of futuristic alchemist's cave.

In the early 1895 time frame, creating one million voltages through his conical coil, Tesla had made the decision to transmit a signal

New York state's Big Apple, New York to West Point, fifty miles away in New York state.

On the 13th day of March 1895, Tesla lost the fruits of his experiments and equipment in a massive fire that destroyed the house that was his lab.

In a fire that almost totally destroyed the six-story basement building located at 33 and 35 South Fifth avenue, this city on the 13th of March on the 13th of March Nikola Tesla, the electrician lost all the equipment he used to been conducting his research. He was on the fourth floor. The floor gave way the apparatus was lowered to the second floor and was laying in a shambles ruin. The apparatus was not insured.

20 March 1895 Electrical Engineer Magazine

The Electrical Review reported he was "utterly depressed with a broken spirit'.

I created a fully-functional machine that was capable of performing many different operations, but the conclusion of my efforts was put off until 1897.

I'm in too much sorrow to even talk. What can I say? The effort of half my life was burned in a fire which took only one half an hour... All has been destroyed.

I was so depressed and down during those times that I don't believe I could have made it through with the constant treatment of electricity that I gave myself. The reason is that electricity pumps in the tired body what it requires most - life energy.

In the meantime, Guglielmo Markini an aspiring Italian scientist was making progress, working hard in England developing an electronic device to wirelessly transmit messages. For Tesla the timing could be better.

In an effort to take an advantage over Marconi the concerned Tesla wanted to establish a new laboratory in quick succession. Friends raised funds to start the lab.

The melancholy was beginning to diminish and some money to enable him to carry on in the summer of 1895, he took the space for his new venture on two floors in a building located at 46 East Houston Street. He started developing wireless technology to transmit power throughout the world.

In September 1897, Tesla was in a position to file an patent on the fundamental idea of radio technology, however it would be another fifty years before Tesla was recognized with the invention.

While many scientists, influenced through the research of Hertz were working on the use of electromagnetic waves to communicate Hertz's invention is recognized as the most closely matches the system to be created.

He imagined broadcast signals being that were transmitted using a specific frequency of the carrier. An array of antennas would respond to a specific frequency, which would allow the transmission of an advanced signal to be created. Again, Tesla's vision was the basis for the environment we live in today.

Investigating spirituality

Afraid to feel sad about the fire in the lab at times, Tesla developed a profound interest in spiritualism as well as Eastern ideas. His educator Swami Vivekananda, an Indian Hindu monk, wrote in a letter on 13 February 1896:

Mr. Tesla was fascinated to learn about The Vedantic Prana and Akasha and the Kalpas that according to him, are the only theories that modern science has the ability to entertain. Mr Tesla believes that he can prove mathematically that matter and force can be reduced to energy potential. I'm planning to meet him next week to see this latest mathematical proof.

The late nineteenth century, Swami Vivekananda was a pivotal person who introduced Indian philosophical concepts in the form of Vedanta as well as Yoga in Western civilization. Western world. Tesla said:

We're spinning through infinite space at an incredible speed. Everything is turning. Everything is spinning. All around us is energy. There has to be a way to utilize this energy more effectively...

With the power that comes from it, and with every kind of energy gathered with no effort, and from the infinite store humanity will make massive leaps.

The mere thought of these incredible possibilities expands our minds, boosts our faith, and gives us a sense of absolute satisfaction.

Chapter 10: Explorations With East Houston

Street

in 1898 Tesla experimented in the East Houston Street laboratory, with a small mechanical oscillator that was attached by an iron column which was lowered to the foundations of the structure.

He was amazed to discover that by altering his oscillator's frequency it could cause furniture and equipment sway across the floor of his lab. In awe of his discovery and unaware of the implications for his decisions elsewhere.

The oscillator's vibrating vibrations were traveling into the building's foundations , causing an even more significant effect. In the local area, Manhattan police received reports of windows breaking as panicked residents fled their offices and homes, believing that an earthquake was rumbling through the area. In a deeper dive into the accounts, they found that the source of the earthquake originated from the laboratory of an important inventor, and went on to look into the issue.

In the in the meantime, Tesla was becoming aware of the force of the vibrations , and he sensed the experiment was into a spiral that was out of the norm. In his zeal to stop the experiment, he struck the oscillator using an Hammer. As he was doing this two police officers stormed into his lab. In a calm manner, Tesla turned and said:

Gentlemen, I am sorry. It's only a little too late to see my experiment. It was necessary for me to stop it abruptly unexpectedly and in a unique manner.

Tesla's fascination with frequency and resonance was fueled by the belief of being able to select the appropriate frequency for the earth's own resonance it could be a way to move the entire world.

Marconi was meanwhile seated with his feet planted on the ground and his attention was focused on the task at hand: the actual process of radio signals being transmitted, which he did at Salisbury Plain, in England. The signal he transmitted reached 5 miles (8km).

Exhibiting Remote Control Boat

After hearing concerning Marconi, Tesla responded with an original invention, which was displayed in a specially-constructed demonstration space, in front of an audience of potential supporters in the Electrical Exhibition in Madison Square Gardens in the month of August 1898. It was just one month after Tesla was granted a second patent, based on the technology he was about reveal.

The awestruck viewers were amazed by the sight of Tesla operating a tiny wooden boat that had no wires but its creator could steer the boat as it went. It was the display of the first radio controlled device anywhere in the world.

The exhibit was held within an indoor water feature. The water was bouncing around the

147

miniature ship that measured four feet long and on the other end, an control box with several levers.

Antennas for receiving signals were extending from the uppermost part of the boat, with the highest one located in the center. Two other antennas had tiny bulb lights on their tips to aid the operator in gauging the direction and the position of the boat under the cover of darkness.

It was powered by a screw propeller of standard size. It was housed in the hull of the boat the battery-powered electric motor controlled the propeller as well as the rudder. It also safeguarded the electronic mechanism to receive the radio signals by Tesla's control box.

With no limitations of a wired connection both the remote device and remote, Tesla's idea could allow users to alter the speed and direction as well as control the onboard features (such as lighting or moving components) even from a vehicle that is moving.

Witnesses to Tesla's experiments frequently confused by the results in the dark, unsure which way to go: laughing or running. In an effort to make his audience feel at ease, Tesla invited onlookers to ask questions of the vessel and check if its "borrowed brain could be able to answer questions. One of the delegates asked "what is cube-root of 64', and the answer was correct, as four light flashes, emitted out of the boat.

To show that there was no one in this boat Tesla needed to open the boat's lid in front of the confused crowd. In an era when radio waves were pioneering and awe-inspiring, it was lesser of a leap to imagine that Tesla could manage the boat by his own mind.

Alongside being intrigued in the concept of remote control as a academic field of study, Tesla also saw the potential of using technology in war--as a way of stop it. In this case, machine would fight machines, not human against machine.

I do not wish that my glory should be based on the invention of a solely destructive weapon... I would rather remember myself as an one who ended the war... That would be my most cherished joy.

When he proposed his solution for military use, the officers believed that the device was too fragile to be able to serve any practical use. There were concerns that the machine could be altered and the signal jammed or susceptible to interference.

Industries and even the US Department of War were not willing to plunge headlong to the near future. The bold assertion that his invention would eliminate all need for warfare rather than the less shrewd idea the device could sink an enemy battleship not make a difference.

In desperate need of funding to fund its research, the perpetually cash-strapped Tesla tried to overcome opposition and developed an instrument that could completely submerged. Tesla also developed a function which made it an slave, able only to respond to commands emanating from the master. This was not enough for people to believe in his idea.

Ever the supporter, after being informed that Tesla was looking into patents abroad, Mark Twain wrote from Austria.

Dear Mr. Tesla,

Are you in possession of obtained the Austrian or English patents for that destructive terror that you've invented and made the war impossible ?--& and if yes, why don't you put a price on them and ask me to market them?

Cabinet ministers I know from both countriesand of Germany as well, and of course William II. In the hotel the last night, a few interested guests were discussing the best way to join forces with the Tsar to disarm. I advised them to look for something more secure than...a fragile paper contract. They should invite the greatest inventors to create something that armies and fleets are powerless and make the war in the future impossible.

I didn't realize that you had already been attending to this... I am sure you're an extremely busy person, but would you take the time to write me a note.

Sincerely,

Mark Twain.

The project was not to go ahead. Tesla moved to a new project.

Moving to Colorado Springs

Thirty-three years old, was he imagining the invention he thought would revolutionize the world. In addition to his four million-volt coil and aspired to boost the voltage to allow wireless messages across Pike's Peak to Paris for the France's Exposition of 1900.

The apparatus was too large for the New York laboratory, and the inventor relocated into Colorado Springs in the summer of 1899 to start an array of hidden investigations. The community of Colorado Springs was in awe as loads of strange-looking special equipment arrived with the mysterious inventor. Tesla entered an Alta Vista hotel room 207 (a number that can be divided 3).

The venture was financed by the company's founder John Jacob Astor IV that Astor was going to create his own cold light technology, which he believed is superior to Edison bulb. Since millions of bulbs sold every year this was a huge market to explore.

Local media were intrigued by his plans and inquired whether he intended to communicate between peak and peak. Tesla, who was proud of his work, Tesla responded that the plan was not "essential stunts". His plans were much more than that.

I concluded that it is feasible, with a very low elevation to send electrical power through the upper air.

Once the circuit was in place, Tesla had no intention of making a profit from a object like fluorescent lighting. He came up with a bold idea to make use of high voltages that could carry electric currents across vast distances, through electric currents across the earth's surface.

The first step was to set up his laboratory on vacant land located close to the northeast from Colorado Springs, at a peak known as Knob Hill. It was situated between and the Colorado School for the Deaf and Blind offering support to families of children as well as those who are in convalescence at the Union Printers Home for local people suffering from tuberculosis as well as black lung.

The floor was a huge solid box that had roofing panels that slid down. In the middle, was an elongated tower that measured 8 feet (24m). To the top was a needle-shaped metal mast that measured one hundred and twenty-two feet (37m) and capped by an iron ball that was 3 inches (1m) across.

The fence around the perimeter was the white flags with the words: "KEEP OUT-- GREAT DANGER!" To dissuade those who strayed from the warning, there was a famous quote in Italian was hung at the entrance to the bizarre wooden structure. It read 'lasciate, ogne, speranza voi ch'intrate'or abandon hope for all who walk through this place. It was a passage from Dante's The Inferno. It was the text on the gates to hell.

As construction progressed, Tesla busied himself studying lightning and the ways in which electricity can be transferred from one location to the next. Tesla studied how air transforms into plasma and then becomes conductive for a short time.

Tesla started the construction of the biggest coil ever built to the present. It was massive, measuring five feet (15m) in diameter . It was a prelude for the magnifying transmitter that he was to build in Wardenclyffe Tower. The whole installation on the mountain's peak was actually a device to produce lightning.

It was finally time to test the gadget.

The test was the first.

In the late evening, Tesla signalled for an assistant to flip a large switch. The electricity emitted from the coil, and danced across the station.

The apocalyptic explosions roared out of the sky with a flurry of synthetic lightning that were that were more than 100 feet tall (30m) came out of the copper ball that lies at the apex of the antenna.

A distance of twenty miles (32k) In the mine town Cripple Creek, the thunder of the experiment was evidently audible.

At once, the chaos stopped. Below the city, Colorado Springs was plunged into darkness. The lightning went out, and the thunder stopped.

Tesla's test of the device caused a fire to the local power company's generator which destroyed it to the anger of residents of the area. They began to be concerned about the consequences of living in a house with Tesla's mysterious master of science.

Unfazed, Tesla spent the next six months pursuing his venture into the field of wireless power.

He set up a test in which he put some bulbs on a ground and then they were enclosed in the fifty-foot (15m) area of wire. He distributed power in such a way the electrical fields was created inside the wire, which lit the bulbs. This led to Tesla's belief in his ability to be able to achieve this on a large size.

At the conclusion of eight months of study, Tesla declared that he was able to transmit electricity in a large amount and economically all over the globe. However Tesla did not provide any sufficient evidence to prove his point. He believed that he had created Earth resonance that, according to his theory can be achieved regardless of distance. Tesla was one of those who, when believed he had evidence it was a matter of speed to promote the idea, and then explain it all over the world. He was not one to be questioned about his findings in full.

In another experiment, Tesla noticed a curious sound that was being received at night, by his equipment, and believed, to his own delight, to an extra-terrestrial communications. Tesla expressed his opinion in a written letter addressed sent to American Red Cross.

Brothers We have received an email from a different universe, a world that is undiscovered and distant. It is a message that reads: one, two, three.

The press broke into a rage. Tesla was slammed for entertaining his "message to his fellow Martians.'

From from the Martians.'

The Colorado Springs Gazette poked fun at him over the discovery, and wrote:

If there are any people on Mars that are a part of Mars, they surely displayed the best taste by picking Colorado Springs as the particular location... to be able to make it possible to communicate.

It's a common standard in science that says "when you're told to make a statement, make it an excellent one'. men have achieved greatness through adhering to this rule.

Colorado Springs Gazette

Though he was frequently ridiculed for his claims, Tesla may have been the first to recognize radio waves coming from space. In 1996, scientists released an article that reconstructed Tesla's experiment and concluding that the radio waves were actually caused by the moon Io traversing the magnetic field of Jupiter.

They're not alien civilizations, but originate from the sun and the stars.

Dennis Papadopoulos

University of Maryland

The final journal entry of his in Pike's Peak was on the 7th January 1900, suggesting improvements he would like to add to the oscillator.

Though I haven't yet been able to transmit a transfer of significant amounts of energy, which could be important for industrial purposes in the vastness using this method... it is clear that the efficiency of the method has been extensively demonstrated.

The return to New York

It was the time that Knob Hill station shut down. Tesla was back in New York, five years after the conclusion of their Niagara project. A hundred thousand dollars of capital was lost during the eight months of testing the idea with Colorado Springs. Then, he was searching at hundreds of millions of dollars more. The financial problems of Colorado were on his way. The year 1904 was the time he was suing for unpaid debts, and the lab was ripped down in the following year. A few years later, the contents were sold at the courthouse to pay off his obligations.

Despite the difficulties, Tesla remained optimistic. It was the beginning of a brand new century. Electricity was the engine that drove the massive expansion of the city. the power supply was affordable and at the fingertips of almost everyone. The subway was built and lit up by the power provided by Tesla's generators, which submerged into the water several hundred miles distant.

Marconi was the founder of his own company, the Marconi Wireless Telegraph Company of America He also landed at New York in 1900, looking to draw investors. After securing British patents, he applied for the US patent to patent his wireless telegraphy device However, the application was rejected due to the fact that the technology proposed was too similar to Tesla's earlier invention.

The two systems overlapped due to Marconi along with other scientists at the time recognized the effectiveness and potential that was this Tesla system. Electromagnetic waves are more for practical use.

In awe of the transformations in electrification that were sweeping New York, a confident Tesla was looking to the future.

While residing in the prestigious Waldorf Astoria hotel, he wrote an outstanding piece for Century magazine.

The first question is where comes the motivational power? What is the source that powers everything? We watch the ocean change its course and streams of rivers flow; rain, wind and snow pound against our windows, steamers and trains come and depart and we hear the rumbling sounds of cars and the voices of the street and we can feel, smell and taste, and we are thinking about all this. All this motion that is happening, from the surge of the vast ocean to the subtle movement that we see with our thoughts, all have only one common root. All the energy comes from a single source and one source that is the sun. It is the sun that is the source of energy that powers all. The sun sustains all human life and provides the human race with energy.

The author described an innovative method to harness the sun's energy using an antenna.

It will be possible to regulate the weather by using electricity and bring together all nations He proposed the creation of a worldwide wireless communications system. If wireless technology is widely used and the earth is transformed into a giant brain that can respond to all its parts.

This reveals more about his vision of power and communication. In a later interview in 1926, with Collier's magazine, his vision was clearly and loudly.

We'll be able communicate instantly with each other, regardless of distance. In addition, by using telephony and television we will be able to hear and see each other as if we were talking one-on-one even over thousands of miles. And the devices that will be able to achieve this will be extremely easy compared to our current phone. One is able to carry a phone in his pocket.

The status of Tesla as a scientist had been significantly reduced at the time. In the past the inventor had not produced an invention that was commercially successful. Finding people willing to fund his work was becoming difficult. Investors who were wary had seen many other investors have their money evaporated through Tesla's experiments and had nothing to show for it.

The Wardenclyffe Facility

Fortunately for Tesla one person was still fascinated by the concept of a global communication network one of the most influential people John Pierpont Morgan (17 April 1837-31 March 1913).

JP Morgan was a US banker and financier who controlled Wall Street throughout what was known as the "Golden Age' that lasted between the 1870s and the beginning of 1900s. In this period JP Morgan was the driving force for the industrialization boom across America.

After attempting to obtain money from powerful business people, Tesla eventually convinced Morgan that he could develop an inter-Atlantic wireless communications system which would be superior to Marconi's radio wave-based short-range Telegraph system.

Tesla's breakthrough that used alternating currents could be worth millions. Morgan wanted to know what the eccentric inventor had to say about his.

Tesla performed a mental trick by suggesting and kept it a secret to transmit electricity in unlimited quantities to anyone with an antenna that was suitable. He believed this would contribute for the benefit of all humanity, however, he was aware that commercially, he could not offer this idea to Morgan. He was a believer in communication of data wirelessly.

In the month of March, 1901, Morgan agreed to inject one hundred 50 thousand dollars in the construction which is around four points five million in modern dollars. After the project was completed, Morgan will have a fifty percent 100 control over the patents. The idea was to develop something that could beat the transatlantic telephone cable. Morgan advised that this investment was not his maximum. (Tesla would require an additional amount of money for the project to be completed and it is believed that more funding would come after he'd achieved the possibility of the success.)

It was only a matter of time before the ink out on this contract, Tesla determined to make his solution more competitive. He wanted to implement his concepts for wireless power transmission over the air. This would require a massive expansion of the operation. The goal was to transmit communications across in the Atlantic to England and also ships in the sea, following Tesla's theories about making use of the Earth to carry the signals. A bold move was made that included facsimile images and telephony would be incorporated. The change was not communicated to his investor prior to the start of work.

The construction was started during the summer of 1900 in the summer of 1900, when Tesla established the operations at Shoreham, Long Island, and later, the creation of a plant that had massive transmission towers called Wardenclyffe which was named for the former owner.

Built entirely from wooden beams, atop of the level, the structure increased up to one hundred eighty-seven feet (57m). The tower illuminated the night sky, and then exploded with a roaring sound, again to the horror of residents in the area.

In the meantime, Marconi's device was undergoing an amazing test. It was at Signal Hill, in modern Newfoundland, Canada, on 12

December 1901 The Italian as well as his colleague, George Kemp, confirmed the reception of the first transatlantic radio signals with a telephone receiver as well as an antenna made of wire and held up by the kite.

It is believed that the Morse codes for the letters 'S' was believed to have been sent by Poldhu, Cornwall, England approximately two thousand two hundred miles away. (Radiographers are still debating whether Marconi was not entirely correct by what he heard because he was aware that three clicks, and the conditions in which the tests were conducted suggested failure, which leaves those who participated to question the authenticity of the outcome.) However, a lot of people celebrated it as a huge achievement at the time that would change investors' minds.

Tesla dismissed the achievement of the Italian.

Marconi is a nice guy. Let him go on. He's using 17 Patents I have issued.

Technical issues were starting to catch up to Tesla and his larger-than-life creation. The funds he had accumulated were in decline.

The demise of Wardenclyffe

As Marconi's achievements became apparent, Morgan began to doubt the worthiness in his decision. Marconi's method not only worked however, it was also cost-effective. After the first test, Marconi had conducted a number of experiments to verify the efficacy of his method.

Morgan requested Tesla to provide an update on how development was progressing'. Because of the drastic deviation from the consensus the project was not progressing in any way.

Tesla had to reveal to JP Morgan his real plan.

What I am thinking about and what I am able to do, Mr. Morgan isn't just a simple transmitting of messages, but rather the global transmission of electricity without wires across vast distances however, it's changing the whole world into a conscious being, or a living thing capable of feeling all its parts , and through which thought can be transmitted through your mind... One power plant with around one hundred horsepower is capable of operating many millions of devices... Are you willing to assist me or let my amazing work, which is almost done, go to the pots?

Morgan was a businessman who was pragmatic and had already made a decision to support Marconi. He was aware of Tesla's modifications and the demand for additional sums of money needed to make it into a breach of contract. He decided to not fund the modifications.

I have received your email and, in response, I would like to say that I'm not in the present time to pursue any further proposals

With the birds of prey

The public was unaware that Tesla was in a state of financial ruin and only saw the successful billionaires' firms that benefited from his innovations.

He began to sink into a state of utter isolation. The world of business to him was uninteresting to him. He struggled to understand the connection between the value of innovation and the significance of business. He didn't understand how to approach his invention from the perspective of a finance professional. One of the most crucial aspects of being creative is knowing how to integrate into the real world in real life, but it didn't come through for him repeatedly. This was a major defect that ruined living a full and satisfying life.

However, Tesla dreamed of resurrecting his dream of Wardenclyffe and fought for years to raise necessary funds.

I'll be able transmit any amount of energy to any point.

In 1916, as his fortune declining in 1916, he gave up in 1916 the Wardenclyffe mortgage in 1916 to Waldorf Astoria Hotel where he was living for nearly twenty years in credit, accruing the amount in the amount of 20 thousand. 535 2019 dollars.

The next year, the Waldorf saw the tower be set on fire to allow the land to be easier to sell.

He was so quiet that he turned to animals for companionship. A few years later the inventor Dragislav Petkovic, who was visiting from Yugoslavia was able to walk with the inventor around Bryant Park on his daily mission of mercy, and come in aid to pigeons the winter months.

Mr. Tesla was looking up at the library window, and noticed that the windows are secured by iron bars so that pigeons could not slide down and then freeze during the nightat night.

In one of the corners the cat was spotted. It was frozen halfway. He advised me to stay there and make sure the cat doesn't try to get him, and I look around to see if there are others'. When I was looking around, I attempted to get the pigeon's attention, but was unable to do so because the bars were too close to each other. As soon as Mr. Tesla came back, he immediately bent the bar and pulled it out. "All childhood memories remain very important to me' the man explained, and he began to smack the frozen pigeon insuring it that it would soon recover. He then took the box from my hand and began throwing the food in the front the building. As he handed out the food the food, he told me: "These are my dear friends.'

The agony of nobel prizes

The year 1909 was the one when Marconi received a Nobel Prize stealing Tesla's limelight again. His sadness turned to anger that lasted for a long time. After 18 years the man loosened his guard and spoke in a more scathing manner than normally:

Mr. Marconi is the name of a donkey.

A lawsuit was filed against Marconi by an unemployed Tesla and Tesla's patent infringement claim. But, a poor man trying to fight an legal battle against an enterprise with deep pockets to pay for their legal representation was in danger. Tesla gave up.

Marconi was awarded his Nobel Prize for work that Tesla was sure to believe was his own. He regretted:

I guess that all is fair in war and love.

It was expected that the Nobel Prize would revisit the life of Tesla in 1915. It was accompanied by a headlines in The New York

Times that the Swedish government was set to give two Nobel Prizes the following week in physics to Thomas A. Edison and Nikola Tesla. He said:

The honor bestowed to me is related to the transmission of electricity without wires.

The award did not awarded. The next week, it was awarded jointly to an English scientist William Bragg of Oxford, England and his son Laurence for their groundbreaking research into crystal structures using X-rays.

The scandalous mix-up between Tesla and Tesla is not yet solved. Two of the most plausible theories is that this was caused by a accidental press release that was mishandled. A rumour also spread that Tesla refused to share the prize with Edison, to make sure he didn't get access to the twenty-thousand-dollar prize fund.

It was a make or break for Tesla. The state of his finances quickly came to light through a few testimonies.

Nikola Tesla, the electrical inventor, was the subject of a lawsuit for $935 back tax was issued today. Mr. Tesla stated under oath his finances were in shambles and he had been living off credit.

To help in their efforts, helping, American Institute of Electrical Engineers gave Tesla their highly coveted Edison Medal, to give the long-overdue honor. At the time of the award, Edison was conveniently away working, while Tesla was not present, preferring to be on the other side of the street, feeding the pigeons in the vicinity of the New York Public Library.

In the later evening, he was accepted the award. Incredibly encouraged, Tesla announced that his invention of transmitting wireless power was now complete.

The energy is transferred into a faraway place and you'll be able to see something similar to an Aurora Borealis. To sum up that, we are achieving fantastic results.

The audience was puzzled as to whether Tesla who was still an eccentric inventor, was now losing his way. When he was asked about his involvement in the Nobel Prize debacle, he was insistent.

I am not willing to give those who are naive and jealous the pleasure of having defied my efforts. They have nothing more than the microbes that carry a dreadful disease.

At this point the man was able to care for injured birds in his bedroom and was still able to feel more connection with them than other human beings. He was getting the generosity of people he knew as well as the generosity of local hotels.

Finally, after securing the odd job as an engineer and a consultant, he was able to pay for a small laboratory and office located in the Metropolitan Tower.

Looking forward to the home run

Today, aged 61, Tesla would spend the remaining years of his life hoping that someone would invest in his inventive inventions he continued to bring forth. In between, he had conversations with a variety of companies, feeling unhappy and disappointed that his ideas never were able to be put into the market.

His primary goal was to create some kind of tangible asset that he could trade to finance the rest of the work required to finish his Wardenclyffe baby -- a solid system that could transmit images, light, and electricity to all corners around the world. Tesla nevertheless managed to draw the attention of people, but his ideas began to lose their realism and his thoughts were slipping more towards science more than actual science, with his attempts to capture thoughts as an example.

In debt, Tesla was able to move from hotel to hotel leaving a trail of debt behind. In 1921 Tesla began inviting birds to return to his hotel room in his St. Regis hotel. The guests were able to leave and come back at their own pace and he would provide nests for pigeons on open windows.

Large flocks of them would flit through his windows and in the rooms. Their scum on the exterior of the building would become an issue for the managers as well as on the inside for the housekeepers.

One of his most loved birdies died, he wrote:

I saw a flash of light in her eyes that was more intense than even the strongest bulbs in my lab. When the light was gone, a glow was released within me. To that point, I was certain that I'd finish my task regardless of how challenging my plan however, when that thing was gone from my life, I knew that my work was over.

Tesla did manage to pay for packages that were sent packages to Mark Twain at a non-existent New York City address, which is even more interesting given the fact that Twain died.

Infrequent visits in Times Square to see the films on the silver screen he could see his futuristic visions pop up. In the meantime the universe of science fiction books and magazines is the only place for his ideas.

Chapter 11: Nikola Tesla's Failings And

Death, And The Legacy That He Left Behind

Following the success in Niagara Falls, following the successful completion of Niagara Falls project, Tesla continued to experiment in his lab located in New York City, particularly with high-frequency electricity. There have been many developments in this field, Tesla wanted to discover an instrument that could allow him to explore uncharted territory.

5.1 Tesla Coil

He began by building rotating alternating current generators that are faster to run. The idea was not successful because the machine breaks down when it gets to 20000 seconds of cycles. The solution was found through the device known as the Tesla coil which was patentable in 1891. With this device, a normal cycle that is 60 times per second could be amplified to extreme frequency of hundreds of times per second. In addition to the high frequencies and the incredibly high frequency, the Tesla coil can also produce massively high voltages. With the high frequencies it was able to generate the first neon and fluorescent illuminations and also capture the first xray images. When he discovered the Tesla coil was the beginning of his fascination for wireless energy transmission.

Through his coils Tesla discovered that he could detect and transmit radio signals with power when they're tuned at the same frequency. In 1985, when he was preparing the test of his gadget and then transmitting his first radio signal the device was destroyed by a fire in Tesla's lab and all of his work. In addition, Guglielmo Marconi, an Italian inventor from England was developing a wireless communication system. In 1986 Marconi was able get a patent on his invention in England. The next time, Tesla sought a patent for his radio. It was granted three years.

In 1900, the year that the Tesla patent was awarded Marconi's wireless company started to gain momentum in the stock market mostly due to his connections with English Aristocracy. From $3, the company's stock rose to $22 a share, and Marconi gained international recognition. On the 12th of

December in 1901, he transmitted messages over the Atlantic Ocean for the first time. While Marconi was making strides in radio technology, Tesla was not fazed as he was aware that Marconi used his patents to create the technology. In 1904, Tesla's faith was shaken after it was discovered that the United States Patent Office presented the patent for the invention in radio Marconi with no explanation the reason. It was only a couple of years after that, in 1911, that Tesla was at his breaking point following Marconi was awarded the Nobel Prize. Tesla filed a lawsuit to redress the infringement against Marconi's business but was unable to afford the funds to pursue lawsuits. Then, he decided to decided to let it go.

5.2 The Wardenclyffe Project

After a long stay at Colorado Springs to do research for El Paso Power Company, El Paso Power Company, Tesla returned to New York and wrote an article for Century Magazine. In his article, Tesla provided a thorough account of his plans to harness the sun's energy by using an antenna and controlling the weather using electricity.

After reading the article, J.P. Morgan was enticed by the plan that was proposed by Tesla. When he visited Morgan's house, Tesla showed him a system of wireless communications that could transmit news and stock market updates and military communications, as well as pictures, private messages and even music to anyone in the world , and also pass over telephone messages across oceans. When he heard that, Morgan asked Tesla to construct a power plant and transmission tower, and provided him with $150,000. Though this was a small sum to what Morgan would like to see happen, Tesla took whatever was offered to him and began working. Tesla purchased a piece of land that was called Wardenclyffe in which the tower was planned to be constructed.

The building of the tower progressed slowly it was clear that it would require more than $150,000 in order to complete the tower. But, Morgan did not respond to the demands of Tesla. When Marconi's device sped the"S" in the letter "S" over the Atlantic Ocean on December 12 the year 1901 Tesla promised Morgan he did not have anything to worry about since Marconi used seventeen of Tesla's patents in order to make this transmission. After seeing Marconi's cheap device be successful, Morgan started to doubt Tesla. Despite the requests of Tesla to offer additional funding, Morgan refused and, the project was ultimately scuttled because of the lack of financial support.

5.3 The death of Nikola Tesla

In 1905 the project was abandoned completely. Tesla was attacked and called the Wardenclyffe Project to be a "million-dollar foolishness". In the aftermath of the humiliation that the project caused him, Tesla went through another emotional breakdown. Then, Marconi was awarded a Nobel Prize and was dubbed the "father of radio" and became wealthy, while Tesla was left in debt. Following this, Tesla tried to work on a handful of projects to keep his reputation. He initially tried to develop turbine engines, but stopped in the end. He also developed the idea of detecting ships in the ocean. His idea was to transmit high-frequency radio waves which could be seen on a screen that was fluorescent. It was not a technology that anyone could take up, yet it was the basis of what's now known as the radio.

In 1912 Tesla was beginning to separate himself a society that was unfair to him. Tesla was obsessed with maintaining cleanliness. He also had a recurrent obsession on the number 3; He began to do everything in three sets. He used to track his steps every time you walked, and required exactly 18 napkins set at the table throughout his meals. Furthermore, he was very sensitive to sounds , and had increased perception of sight.

In the last years during his lifetime, the man was obsessed with pigeons, especially white females. He adored the birds as humans. According to Tesla the story, a white pigeon came to his window in the evening to inform him that he was about to die. The pigeon then died within his arms Tesla said it was an indication that his career has ended.

Conclusion

Nikola Tesla is undoubtedly one of the most renowned and famous engineers of the past. He could alter the course of the world through his creation of the magnetic rotating field as well as the system of alternating current. His concept set the foundation for the development of transmission, and the use of electricity in the present. The other contributions he made to science are equally significant. When he died, death, he had more than 1000 patents, including important inventions, like the vacuum tubes that were wireless and the wireless transmission of energy, the basic RADAR as well as laser technologies, the modern electric motors lighting, fluorescent and neon as well as air-friction speedometers remote control, xray images, and the Tesla coil that is utilized in radio, television as well as others electronic equipment. He sparked a wave of technological innovation in the latter part of the 19th century and the early 20th century. He is still inspiring and motivate the current engineers who admire his work.

Sadly, his struggles as well as the challenges he encountered led to his failures , and eventually his fall. Although he had a myriad of ideas in his mind however, his financial issues caused him to be unable to see these concepts to become reality. His ignorance of the business and finance world allowed other people to profit from his inexperience and exploit Tesla's ideas to their own advantage. Future visionaries must learn from his mistakes They shouldn't believe in others too fast or neglect the significance of finance. Innovation and intelligence by themselves aren't enough to ensure success.

Today Nikola Tesla's name has been acknowledged with respect and admiration, which he deservedly merits. Many scientific and engineering institutions and organizations use Tesla as an example of excellence. He was adept at sharing his knowledge with the world . the world needed be benefited by his ideas. Unfortunately, his constantly working mind that generated all of his ideas also caused a lot of stress on his health. At the end of his life his death, he was left homeless and without a partner, leaving behind only his legacy that he'd worked to create throughout his life.

CPSIA information can be obtained
at www.ICGtesting.com
Printed in the USA
BVHW051252080223
657977BV00015BA/751

9 781774 855572